ColdFusion Lists, Arrays, and Structures

Jeff Peters

Proton Arts – Manassas VA USA

ColdFusion Lists, Arrays and Structures

SECOND EDITION: June 2007

01 00 09 08 07 8 7 6 5 4 3 2

ISBN 0-9752647-8-8

Interpretation of the printing code: The rightmost double-digit number is the year of the book's publication; the rightmost single-digit number is the number of the book's publication. For example, the printing code 04-1 shows that the first publication of the book occurred in 2004.
Published in the United States of America

Trademarks

All terms mentioned in this book that known to be trademarks or service marks have been appropriately capitalized. Proton Arts cannot attest to the accuracy of this information. Use of a term in this book should not be regarded as affecting the validity of any trademark or service mark. ColdFusion is a registered trademark of Adobe, Inc. Fusebox is a registered trademark of Fusebox, Inc. BlueDragon is a trademark of New Atlanta Communications.

Warning and Disclaimer

This book is designed to provide information about ColdFusion programming. Every effort has been made to make this book as complete with respect to its topic and as accurate as possible, but no warranty of fitness is implied.
The information is provided on an as-is basis. The author and Proton Arts shall have neither liability nor responsibility to any person or entity with respect to any loss or damages arising from the information contained in this book or from the use of the programs that may accompany it.

This book was written and set with OpenOffice.org 2.2.0

Dedication

For my grandfather

There are only a few remaining in the world who know what the job title "pattern maker" implies. At one time it was a title of craftsmanship, precision, care and respect. It is my fervent hope that in some way the work I do might reflect the pattern he was.

– Jeff Peters

About the Author

Jeff Peters is a project manager and application architect for NCI Information Systems, Inc. Jeff has been involved in information systems in one way or another since 1973, when he first crash-landed a lunar module on a teletype terminal to the Dartmouth University mainframe. That same machine predicted that he would one day be either a shepherd or President of the United States. Neither of these seems to have been fulfilled.

Involved in the Internet since the late '80s, Jeff was unfortunately working on mainframe projects (taking a hiatus from LAN-based systems) when the 'Net went public and changed the face of application delivery forever. Some days he thinks he still hasn't caught up, but he's found that writing books is a good way to keep in practice. Jeff has written or co-authored ten books on web development and writing for techies, and has contributed to publications such as ColdFusion Developer's Journal and PocketPC Magazine. His web-development-oriented web site, GrokFusebox.com, provides a gateway to a wide variety of resources. Jeff can be reached at jeff@grokfusebox.com.

Acknowledgments

I'd like to thank and blame Hal Helms for an awful lot of things. As for many, Hal was the impetus for my involvement in Fusebox, and the goad that started me writing for CFDJ and subsequently venturing into books. Fortunately, Hal is at least as much action as he is talk, and so has not only pushed me along but pulled as well, lending his time as sounding board and occasional editor.

This second edition requires thanks to our readers who have pointed out errata and made suggestions for another edition. Among these are Ben Forta, Sean Corfield, John Beynon, James Husum, and Rey Muradez. I'd also like to thank Adobe for continuing to improve ColdFusion and make volumes like this necessary. It is gratifying to know that the many predictions of CF's demise have been hasty to say the least. Thanks to Gavin Baumanis, whose eagle eye helped spot problems all the way from Down Under. Good on ya, mate.

My wife, Kathryn, remains ever the patient and understanding partner in my literary crimes. It can only be providence that she is a writer probably even more than I. I also want to thank Elizabeth, my daughter whose perspective on the world often makes me stop and rethink things. Oh, to see through the eyes of a child again!

Thanks are also due to you, the reader. I appreciate the fact that you've risked your hard-earned money to purchase this book, and hope that you will find more than the cover price worth of value in its pages.

I save for last and most important my thanks to my heavenly Father, maker of heaven and earth, and his son Jesus Christ. The assembly of words into books or electrons into programs do not do justice to the word "creativity".

Jeff Peters
June, 2007 (updated)

Table of Contents

1. Introduction

Welcome to ColdFusion Lists, Arrays, and Structures. This book is designed to be a part of your ColdFusion coding toolbox. Just as master craftsmen rely on a lifetime's knowledge of their tools and materials to create masterpieces, so do you rely on your box of master coder's tools to create superior programs. This volume provides in-depth knowledge of ColdFusion's complex data types.

The book is organized very simply, with a chapter for each of the major areas: Lists, Array, and Structures. I don't go into recordsets or ColdFusion Components (CFCs). Though these are both complex data types in CF, they are not included because recordsets are very similar in behavior to structures, and CFCs deserve treatment all by themselves. You can find such a treatment in Discovering ColdFusion Components (Helms and Quarto-von Tivadar, Techspedition 2003).

I'll be using plenty of examples in these main chapters to help you find your way through the narration. After the main chapters, I'll get into some of the more adventurous topics related to complex data types, including taking advantage of WDDX and Javascript in your ColdFusion development. While this is not intended to be an exhaustive reference on these topics I hope to at least pique your interest.

Above all, this book is intended to be thought-provoking and fun. It is intended to be kept close at hand while you work. I strongly encourage you to do so. Play with the examples, see how you might adapt them to ideas you're having as you go through the book. Though this is a good reference that you can pull out when you can't remember the details of a particular method; it's also a toy for you to play with when you're in the mood to take things apart and put them back together , figuring it out as you go. I cover some serious topics, but I hope it's never so serious that you don't feel like playing with it.

Now you know what the book is about and why I wrote it; let's take a quick look at who I think should buy this book.

Who Should Buy This Book

Though it's incredibly tempting to say, "Everyone should buy this book, even if they don't know ColdFusion and aren't involved in web development," there really are some folks in particular who I think will find this a volume particularly valuable.

The first sort of person I see reading this book is the ColdFusion developer who wants to explore the language. Often we get so involved in the day-to-day work of turning out applications that we don't get time to fully explore the capabilities of the language we use every day. This is true of practically every developer, regardless of the language or platform they use. When you have a few minutes that might be used for exploration, it's hard to know where to start to best take advantage of those few minutes. You can just grab this book and pick an example to play around with. No single concept takes very long to get a grasp on, so you can rapidly bolster your understanding of ColdFusion in small chunks scattered around your shedule.

The next sort of person I see reading this book is the developer who is preparing for certification exams. Now, don't take that statement wrong; I have no relationship with any of the exam designers, nor can I make any kind of guarantees about improving exam scores if you buy this book. However, I do know that the more comfortable you are with a subject, the more likely you are to test well. This book is definitely about comfort. If complex data types make you nervous, by the time you're done with the book, they'll be relegated to a time-share with the monsters in the closet. If you are fairly familiar with complex data types, you'll be boldly going wherever you care to go.

This brings us to the next sort of person who should buy this book: the explorer--the developer who is always looking for new techniques, always pushing the envelope. This book will stimulate your adventurous tendencies and (hopefully) show you ways to think about complex data that you haven't already considered. This book is not about doing things the way everybody else does. This book is about knowing complex data manipulation so comfortably that you'll be the inventor on your development team, the person who shows the others cool new ways of doing things.

Finally, if you're a ColdFusion developer who just wants to understand CF's complex data types, this is the book for you. You can jump directly to the area of interest and go from there.

Whatever your experience level with ColdFusion, this book offers food for thought, and some fun recipes as well. So let's get to it and start off with a few definitions to make sure we're all speaking the same language.

Definitions

There are really only three definitions I want to make clear before we get right into the guts of the book. If you're an experienced developer, you're already familiar with these concepts. If you're new to ColdFusion a quick stop here will save some confusion later on.

Simple Data Types

Simple data types contain single discrete values, such as a person's e-mail address, their birthdate, or their age. The simple data types in ColdFusion are:

String - can contain any characters

Number - can contain only digits and (optionally) a decimal character

Datetime - can contain only a datetime value (really just a specialized number)

Boolean - can contain only True (Yes) or False (No)

Simple data types are the building blocks of data manipulation within ColdFusion. When we need to start manipulating more than single values, we need to consider complex data types.

Complex Data Types

Complex data types contain multiple values, and advanced organization of those values. Languages that allow complex data types generally offer built-in means to manipulate the values within complex data types. ColdFusion provides a strong library of functions for manipulating each of its complex data types. The complex data types in ColdFusion are:

List - contains a list of values, delimited typically (but not necessarily) by commas.

Array - contains a set of values referenced by index numbers.

Structure - contains a set of values referenced by key names. Structures are also known as associative arrays.

Recordset - contains a set of values arranged in row/column fashion. Recordsets are powerful constructs in ColdFusion, but not part of the scope of this book.

ColdFusion Component (CFC) – CFCs resemble structures in the way they store data, but they also have methods (behaviors) in addition to data. Although CFCs are technically a complex data type, they are beyond the scope of this book.

Looping

Complex data types, by definition, may contain multiple values. Therefore, it is often necessary to perform the same action on each value within a complex variable. This practice is known as looping, and ColdFusion provides a number of ways to create looping constructs. The most prevalent of these relative to lists, arrays, and structures is the **`<cfloop>`** tag. You'll be seeing this tag quite a bit throughout the book. When further exploring the use of complex data types, you should also be aware that looping constructs can also be created with the **`<cfoutput>`** and **`<cfmail>`** tags. At any rate, the bottom line of looping is that in order to make it successful, we need to create dynamic methods of referring to the various members of a complex data type. The chapters on lists, arrays, and structures will go into these methods in detail.

Finally, let's take a look at an overview of the chapters that will be covered in the book. A quick perusal of the next section will give you a complete overview of the book.

Chapter Summaries

Chapter 1 - Introduction

This is the chapter you're reading right now. It provides an overview of the book and some suggestions on who might be interested in buying it. If you've read this far, you're probably one of those people.

Chapter 2 -Lists

This chapter explores the creation and use of lists. I'll cover building your own lists as well as taking advantage of lists that may be hiding in your applications. As with the other major chapters, there are plenty of examples to whet your appetite.

Chapter 3 - Arrays

This chapter introduces arrays, starting with one-dimensional arrays and progressing to multiple dimensions. I'll look at ColdFusion's powerful collection of functions for manipulating arrays, but more importantly I'll dive into some example that should make you think of arrays in entirely new ways.

Chapter 4 - Structures

Once you have arrays under your belt, structures are a simple variation on the theme. This chapter will have you creating symphonies of structures, manipulating data in perfect harmony.

Chapter 5 - WDDX and Javascript

Web Distributed Data Exchange (WDDX) provides an XML-based means of moving complex data among disparate environments. Two of the strongest uses of WDDX for ColdFusion developers are the ability to move data back and forth between the Javascript and ColdFusion environments, and to store complex data types in scopes that only allow simple data types. This chapter looks at the use of WDDX to take advantage of complex data in the front end as well as the back end of your applications. With the increasing interest in AJAX-related techniques, Javascript is increasingly important.

Chapter 6 - Loose Ends

This chapter is our last chance to have a look at what we've covered, techniques we've implemented, and ideas we've explored. We'll explore some comparative examples using lists, arrays and structures. I'll also provide some references to resources for further exploration.

Conclusion

This book is designed to be your in-hand reference to ColdFusion lists, arrays, and structures. It is not, however, a simple language reference manual. Rather, it provides explanation of the workings of complex data types and ideas about exploring the use of these data types in your applications.

Now that we know which tools we're going to use, let's start working with them. The first tool in the box is the list.

2.Lists

What Are Lists?

A **list** is a set of simple data points. From a technical perspective, it is a complex string. That is, a list is a string that can contain multiple values.

Most of us use lists fairly frequently, often without even paying attention to them. One of the most common examples of a list is the ubiquitous grocery list. I am a big list user. When I go to a store (for me, any store, not just groceries), I do one of two things: I either make a list, or I forget one or more of the items I was supposed to buy.

Think about a shopping list for a moment. Typically, it's a small piece of paper with items written on separate lines. Why do we write lists this way?

I remember the first time I had a spelling test in second grade. That was a long time ago, but the memory is very clear, for reasons you'll see in a moment.

The teacher instructed us to write each word after its number, each following the one before. My page ended up looking something like this:

1. mammal 2. reptile 3. marsupial 4. bear 5. kangaroo
6. giraffe 7. amphibian 8. lemming 9. boar 10. centipede

Of course, the teacher was expecting something that looked more like this:

1. mammal
2. reptile
3. marsupial
4. bear
5. kangaroo
6. giraffe
7. amphibian
8. lemming
9. boar
10. centipede

Clearly, the second format is far easier to read and comprehend than the first. The second example is clearly a list of individual items, arranged one per line. The first example is nowhere near as easy to understand immediately. Do the lines have significance? Why are there four items on the first line and only three on the other two? These are the kinds of questions that might be asked about the first format. It is not as well defined a list as the second.

Where the first format above doesn't have a clear separation between individual items in the list, the second format does. The second format uses line breaks to separate items. Even if we didn't have numbers in front of each word, the line breaks would make the items in the list clear, just as the items on a shopping list are clearly distinct, one from another. This separation of items is called delimiting.

Delimiting allows us to easily operate on the individual elements of the list. In the case of the shopping list, the line break delimiters let us easily cross off items as we pick them up. In much the same way, we use delimiters to separate the values in a ColdFusion list so our programs can easily work with the discrete values contained within the list.

Delimiters

A **delimiter** is a divider between two pieces of data. The default delimiter for a list is the comma. So in ColdFusion, the list from my fateful spelling test would look like this:

```
mammal,reptile,marsupial,bear,kangaroo,giraffe,amphibian,lemming,boar,centip
ede
```

If I wanted to create a variable to store this list, I would just use `<cfset>` to take care of it:

```
<cfset spellingList="mammal,reptile,marsupial,bear,
kangaroo,giraffe,amphibian,lemming,boar,centipede">
```

Notice the quotation marks at the beginning and end of the list. These reflect the fact that a list, as mentioned earlier, is a string-based data type.

With comma delimiters inserted in the list, we can let ColdFusion help us work with the elements of the list. For starters, we've lost the line numbers from the original list. But order is significant where lists are concerned. In other words, the position of a list element can be referenced by an index number. This is done with the ListGetAt() function. So if I wanted to know the fifth word in the list, I could use this:

```
ListGetAt(spellingList,5)
```

to get the value at position 5 of the list. This sample would return "kangaroo". It's a simple matter of counting each element as separate from the next when you run into a delimiter.

ColdFusion has a very strong set of functions that are designed for manipulating lists. There is a complete discussion of these functions later in the chapter. Before we get to the reference materials, though, I'd like to take a look at what I call overt and covert lists.

Overt Lists

Although this sounds like some kind of mystical idea, **overt lists** are just lists where the delimiters are readily recognizable. Our spelling list is a good example. Some others are:

A class role:

```
"Baker,Bartholomew,Bertrand,Caesar,Mitchell,Walker,Zigler"
```

A list of store locations:

```
"Downtown,Midtown,Springfield,City Mall"
```

A numeric sequence:

```
"1,1,2,3,5,8,13,21,34,55,89,144"
```

A developer who has worked with lists would look at any of these examples and instantly recognize it as a list. This is why I call them overt lists. As you can imagine, covert lists are another matter altogether.

Covert Lists

A **covert list** is a set of values that is probably not immediately recognized as a list. We use a large variety of information in our daily lives that falls into this category. The reasons we typically don't recognize a covert list as being a list are that it uses some delimiter other than a comma, and we are usually accustomed to seeing the list as a single chunk of data.

Here are some examples of covert lists, and the delimiters that they use:

A Social Security number (dashes):

```
123-45-6789
```

A telephone number (dashes):

```
703-555-1212
```

A telephone number (dots):

```
703.555.1212
```

A full name (spaces):

```
George W. Bush
```

A floating point number (decimal):

```
3.1416
```

An e-mail address: ('at' symbol or dot):

```
jeff@grokfusebox.com
```

A list can be delimited by just about any character you choose. When you recognize this idea, you gain a powerful tool for manipulating string data. I call it "breaking the comma barrier".

Breaking the Comma Barrier

As the examples in the previous section illustrate, the options for delimiters in lists are numerous. Let's consider that for a minute. If we were only allowed to use the comma as a list delimiter, ColdFusion's list functions would be much less powerful than they are. But because we can specify what character we want ColdFusion to recognize as the delimiter, we can take advantage of the list functions in a vast number of problem areas. All that is required is the ability to "think outside the box". In this case, the tendency to think of lists as being solely comma-delimited is the box. Once we get outside it, the possibilities abound.

We also want to get outside another box--the tendency to use the first function we find that will do the job. For example, many developers discover the ListGetAt() function when they first explore the need to retrieve an element from a list. It works, and so they may never find out about ListFirst() and ListLast(), which are easier ways to represent actions that can also be performed with ListGetAt(). For example, let's say we have a list with four elements and we want to get the first and fourth elements of the list. We could have a block of code that looks like this:

```
<cfset lstBeatles = "John,Paul,George,Ringo">
<cfset firstBeatle = ListGetAt(lstBeatles,1)>
<cfset lastBeatle = ListGetAt(lstBeatles,4)>
<cfoutput>
  The first Beatle is #firstBeatle#,
  and the last Beatle is #lastBeatle#.
</cfoutput>
```

This is fine, and it works.

But ListGetAt() isn't the only way to solve the problem at hand. Have a look at this code as well:

```
<cfset lstBeatles = "John,Paul,George,Ringo">
<cfset firstBeatle = ListFirst(lstBeatles)>
<cfset lastBeatle = ListLast(lstBeatles)>
<cfoutput>
  The first Beatle is #firstBeatle#,
  and the last Beatle is #lastBeatle#.
</cfoutput>
```

This code does the same thing as the first block, with one subtle difference. Instead of explicitly pulling the elements in the first and fourth positions in the list using ListGetAt(), we pull the first and last elements using ListFirst() and ListLast(). The advantage in coding is that we don't need to know the position of the last element (the length of the list), and there's a little less typing to do.

The difference in behavior can be easily seen if the content of the list is altered. The final version of our example is shown below, followed by its output.

```
<cfset lstBeatles = "John,Paul,George,Ringo,Pete">
<cfset firstBeatle = ListFirst(lstBeatles)>
<cfset lastBeatle = ListLast(lstBeatles)>
<cfoutput>
  The first Beatle is #firstBeatle#,
  and the last Beatle is #lastBeatle#.
</cfoutput>
```

The first Beatle is John, and the last Beatle is Pete.

Illustration 1- List Functions Output

So the ListLast() function allowed us to adapt to the new list without changing the code. With a good handle on these basic list functions, we can explore how they might be used in conjunction with covert lists and delimiters other than the comma.

When writing web applications, we often have to deal with email addresses. If we recognize an email address as a covert list, we can perform some useful operations on it. For starters, maybe we have different responses we'd like to use depending on whether a customer is a commercial or educational representative. The following code will extract the top-level domain from the email address:

```
ListLast(email,".")
```

This piece of code shows the technique in action:

```
<cfset address1 = "jeff@grokfusebox.com">
<cfset address2 = "student@ourcollege.edu">
<cfoutput>
  address 1 has a TLD of #ListLast(address1,".")<br>
  address 2 has a TLD of #ListLast(address2,".")
</cfoutput>
```

NOTE: In many cases, particularly with non-North-American domains, the desired domain might be represented by more than one member in the list. For example, an Australian university domain would end in **edu.au**.

Similarly, if we needed to extract a list of the mail servers from which we get mail, we'd need to pull off the tail ends (after the @ symbol) of all the email addresses. This would be trivial using ListLast() in a loop over all the addresses:

```
<cfoutput query="userEmails">
#ListLast(email,"@")#<br>
</cfoutput>
```

Deciding on which function to use is only possible if we take the time to get familiar with all the ColdFusion list functions and how they can be used. That's exactly what we'll do in the next section. After we've looked over all the functions, we'll take a look at some special cases when handling lists, and then plenty of examples.

ColdFusion List Functions

You can find a complete reference to ColdFusion's list functions in the online documentation that comes with CF and CFStudio. This section will take a quick run through this information, so you'll have it handy in this guide just in case you can't get to the online reference. Keep in mind that all of the list functions allow a value to be added for the delimiters argument in cases where you use a delimiter other than the comma. This argument is always the last one in any list function call.

The code samples in this section are intended to be as uniform as possible, to allow learning by direct comparison of different samples and their results. You'll frequently see <cfdump> used to examine the results.

ListAppend() Function

ListAppend() adds a new element to the right end of a list. It returns the list with the new element added.

```
<cfset myList = "Able,Baker,Charlie">
<cfset myList = ListAppend(myList,"Delta")>
<!--- myList now contains "Able, Baker, Charlie, Delta" --->
```

Cool Tip: If you ListAppend() a list to another list, the result is a concatentation of the two lists. In other words, each list is treated as a simple string variable. The following example illustrates this technique.

```
<cfset myFirstList = "Able,Baker,Charlie">
<cfset mySecondList = "One,Two,Three">
<cfset myList = ListAppend(myFirstList, mySecondList)>
<!--- myList contains "Able,Baker,Charlie,One,Two,Three" --->
<cfdump var="#myList#">
```

Note that the two lists are appended together using the delimiter specified by the ListAppend() function, regardless of what delimiter exists in either of the original lists.

```
<cfset myFirstList = "Able:Baker:Charlie">
<cfset mySecondList = "One;Two;Three">
<cfset myList = ListAppend(myFirstList, mySecondList)>
<!--- myList contains "Able:Baker:Charlie,One;Two;Three" because the
ListAppend() function used its default delimiter, the comma. --->
<cfdump var="#myList#">
```

We can also specify the delimiter to use:

```
<cfset myFirstList = "Able:Baker:Charlie">
<cfset mySecondList = "One:Two:Three">
<cfset myList = ListAppend(myFirstList, mySecondList, ":")>
<!--- myList contains "Able:Baker:Charlie:One:Two:Three" because we
specified the colon as the delimiter for the ListAppend() function. --->
<cfdump var="#myList#">
```

ListChangeDelims() Function

ListChangeDelims() allows the delimiter of a list to be changed. It returns the list with the new delimiter(s). This is very useful when combining lists that come from disparate sources. The third parameter specifies the delimiter to replace; the default is a comma.

```
<cfset desiredDelimiter = ":">
<cfset myFirstList = "Red:Green:Blue">
<cfset mySecondList = "Purple,Orange,Chartreuse">
<cfset mySecondList = ListChangeDelims(mySecondList,desiredDelimiter)>
<cfset myList = ListAppend(myFirstList, mySecondList,desiredDelimiter)>
<!--- myList contains "Red:Green:Blue:Purple:Orange:Chartreuse" --->
<cfdump var="#myList#">
```

Here's a quick way to prepare a list of string values for export to a comma-delimited text file format:

```
<cfset desiredDelimiter = """,""">
<cfset myList = "Jeff:Peters:jeff@grokfusebox.com">
<cfset myList = ListChangeDelims(myList,desiredDelimiter,":")>
<cfset myList = """#myList#""">
<!--- myList contains '"Jeff","Peters","jeff@grokfusebox.com"' --->
<cfdump var="#myList#">
```

Note: There's a more refined way to do this using ListQualify().

ListContains() Function

ListContains() determines if a value is a **substring** of any element of the list. It returns the index of the element that matches the search criterion. If the criterion is not found, it returns 0. Note that non-zero values in ColdFusion evaluate as True, and 0 evaluates as False. When using this function, it's important to remember that a substring will match as well as the entire string, and that case is considered ("ace" does not match "Ace", for example).

```
<cfset myList = "AceSpades,KingSpades,QueenSpades,JackSpades,TenSpades">
<cfset isAceHeartsThere = ListContains(myList,"AceHearts")>
<cfset isAceThere = ListContains(myList,"Ace")>
<cfset isLittleAceThere = ListContains(myList,"ace")>
<!--- isAceHeartsThere is False; isAceThere is True; isLittleAceThere is
False --->
<cfdump var="#isAceHeartsThere#">
<cfdump var="#isAceThere#">
<cfdump var="#isLittleAceThere#">
```

ListContainsNoCase() Function

ListContainsNoCase() works the same way as ListContains(), with one important difference. When using ListContainsNoCase(), the case of characters is not considered ("ace" matches "Ace", "ACE", "ace", etc.).

```
<cfset myList = "AceSpades,KingSpades,QueenSpades,JackSpades,TenSpades">
<cfset isAceHeartsThere = ListContainsNoCase(myList,"AceHearts")>
<cfset isAceThere = ListContainsNoCase(myList,"Ace")>
<cfset isLittleAceThere = ListContainsNoCase(myList,"ace")>
<!--- isAceHeartsThere is False; isAceThere is True; isLittleAceThere is
True --->
<cfdump var="#isAceHeartsThere#">
<cfdump var="#isAceThere#">
<cfdump var="#isLittleAceThere#">
```

ListDeleteAt() Function

ListDeleteAt() is the way to remove an element from a specified position in a list. Numbering is left to right, starting with 1.

```
<cfset myList = "AceSpades,KingSpades,QueenSpades,JackSpades,TenSpades">
<cfset myList = ListDeleteAt(myList,3)>
<!--- myList shows "QueenSpades" has been deleted. --->
<cfdump var="#myList#">
```

ListFind() Function

ListFind() is very closely related to ListContains(), with one critical difference. Where ListContains() will match on a substring of an element of the list, ListFind() will only match on the complete element.

```
<cfset myList = "AceSpades,KingSpades,QueenSpades,JackSpades,TenSpades">
<cfset isAceHeartsThere = ListFind(myList,"AceHearts")>
<cfset isAceThere = ListFind(myList,"Ace")>
<cfset isAceSpadesThere = ListFind(myList,"AceSpades")>
<!--- isAceHeartsThere is False; isAceThere is False; isAceSpadesThere is
True --->
<cfdump var="#isAceHeartsThere#">
<cfdump var="#isAceThere#">
<cfdump var="#isAceSpadesThere#">
```

ListFindNoCase() Function

ListFindNoCase() has the same relationship to ListContainsNoCase() that ListFind() has to ListContains(). That is, where ListContainsNoCase() will match on a substring of an element of the list, regardless of case, ListFindNoCase() will only match on the complete element, regardless of case.

```
<cfset myList = "AceSpades,KingSpades,QueenSpades,JackSpades,TenSpades">
<cfset isAceHeartsThere = ListFindNoCase(myList,"AceHearts")>
<cfset isAceThere = ListFindNoCase(myList,"Ace")>
<cfset isAceSpadesThere = ListFindNoCase(myList,"acespades")>
<!--- isAceHeartsThere is False; isAceThere is False; isAceSpadesThere is
True --->
<cfdump var="#isAceHeartsThere#">
<cfdump var="#isAceThere#">
<cfdump var="#isAceSpadesThere#">
```

ListFirst() Function

ListFirst() is one of the simplest and most frequently used of the list functions. It returns the first element of the list.

```
<cfset myEmail="jeff@grokfusebox.com">
<cfset firstMemberPlain = ListFirst(myEmail)>
<cfset firstMemberAt = ListFirst(myEmail,"@")>
<cfset firstMemberDot = ListFirst(myEmail,".")>
<!--- firstMemberPlain is "jeff@grokfusebox.com"; firstMemberAt is "jeff";
firstMemberDot is "jeff@grokfusebox" --->
<cfdump var="#firstMemberPlain#">
<cfdump var="#firstMemberAt#">
<cfdump var="#firstMemberDot#">
```

ListGetAt() Function

ListGetAt() allows us to extract an element from a specified position in the list. Numbering is left to right and starts at 1.

```
<cfset myList = "AceSpades,KingSpades,QueenSpades,JackSpades,TenSpades">
<cfset thirdMember = ListGetAt(myList,3)>
<!--- thirdMember is "QueenSpades" --->
<cfdump var="#thirdMember#">
```

ListInsertAt() Function

ListInsertAt() allows us to add an element to the list at a specified position. The elements of the list starting at the specified position are shifted one position to the right.

```
<cfset myList = "AceSpades,KingSpades,QueenSpades,JackSpades,TenSpades">
<cfset myList = ListInsertAt(myList,3,"Joker")>
<!--- myList shows "Joker" has been inserted in the third position. --->
<cfdump var="#myList#">
```

ListLast() Function

ListLast, like its counterpart ListFirst(), is one of the simplest and most frequently used of the list functions. It returns the last element of the list.

```
<cfset myEmail="jeff@grokfusebox.com">
<cfset lastMemberPlain = ListLast(myEmail)>
<cfset lastMemberAt = ListLast(myEmail,"@")>
<cfset lastMemberDot = ListLast(myEmail,".")>
<!--- lastMemberPlain is "jeff@grokfusebox.com"; lastMemberAt is
"grokfusebox.com"; lastMemberDot is "com" --->
<cfdump var="#lastMemberPlain#">
<cfdump var="#lastMemberAt#">
<cfdump var="#lastMemberDot#">
```

ListLen() Function

ListLen() returns the number of elements of the list.

```
<cfset myList = "AceSpades,KingSpades,QueenSpades,JackSpades,TenSpades">
<cfset cardCount = ListLen(myList)>
<!--- cardCount is 5 --->
<cfdump var="#cardCount#">
```

ListPrepend() Function

ListPrepend() adds an element to the beginning of the list. Lists go from left to right, so the beginning of the list is the left end.

```
<cfset myList = "AceSpades,KingSpades,QueenSpades,JackSpades">
<cfset myList = ListPrepend(myList,"Joker")>
<!--- myList is "Joker,AceSpades,KingSpades,QueenSpades,JackSpades" --->
<cfdump var="#myList#">
```

ListQualify() Function

ListQualify() provides a smart way to wrap elements of a list in qualifiers (usually quotes). This is the better way to make a comma-delimited format out of a list, as we did in the example for ListChangeDelims(). ListQualify() allows the use of an argument to specify whether you want every element of the list to be qualified, or only the elements that have character (non-numeric) data.

```
<cfset myList = "One,2,Three,4,Five">
<cfset myQualifiedList = ListQualify(myList,"'")>
<!--- myQualifiedList is "'One','2','Three','4','Five'" --->
<cfset myCSVList = ListQualify(myList,"'","CHAR")>
<!--- myCSVList is "'One',2,'Three',4,'Five'"; notice the numeric elements
have not been qualified. --->
<cfdump var="#myQualifiedList#"><br>
<cfdump var="#myCSVList#">
```

ListRest() Function

ListRest() is the inverse of ListFirst(); that is, instead of returning the first element of the list, ListRest() removes the first element and returns the rest of the list.

```
<cfset myList = "AceSpades,KingSpades,QueenSpades,JackSpades,TenSpades">
<cfset myList = ListRest(myList)>
<!--- myList is "KingSpades,QueenSpades,JackSpades,TenSpades" --->
<cfdump var="#myList#">
```

ListSetAt() Function

ListSetAt() provides the ability to change the value of the element at a specified position in the list.

```
<cfset myList = "AceSpades,KingSpades,QueenSpades,JackSpades">
<cfset myList = ListSetAt(myList,3,"QueenHearts")>
<!--- myList is "AceSpades,KingSpades,QueenHearts,JackSpades" --->
<cfdump var="#myList#">
```

ListSort() Function

ListSort() sorts the list according to the specified sort type and sort order and returns the sorted list.

```
<cfset myList = "AceSpades,KingSpades,QueenSpades,JackSpades">
<cfset myList = ListSort(myList,"Text")>
<!--- myList is "AceSpades,JackSpades,KingSpades,QueenHearts" --->
<cfdump var="#myList#"><br>
<cfset myList = ListSort(myList,"Text","Asc")>
<!--- myList is "QueenHearts,KingSpades,JackSpades,AceSpades " --->
<cfdump var="#myList#">
```

ListToArray() Function

ListToArray() converts a list to a ColdFusion array and returns the array. We'll discuss arrays at length in the next chapter.

```
<cfset myList = "AceSpades,KingSpades,QueenSpades,JackSpades">
<cfset myArray = ListToArray(myList)>
<!--- myArray is an array whose elements match the elements of myList. --->
<cfdump var="#myList#"><br>
<cfdump var="#myArray#">
```

ListValueCount() Function

ListValueCount() returns the number of elements in the list that exactly match a specified value. It does NOT match substrings.

```
<cfset myList = "AceSpades,KingSpades,QueenSpades,JackSpades">
<cfset spadeCount = ListValueCount(myList,"Spades")>
<cfset kingSpadeCount = ListValueCount(myList,"KingSpades")>
<cfset smallKingCount = ListValueCount(myList,"kingspades")>
<!--- spadeCount is 0, kingSpadeCount is 1, and smallKingCount is 0. --->
<cfdump var="#spadeCount#"><br>
<cfdump var="#kingSpadeCount#"><br>
<cfdump var="#smallKingCount#">
```

ListValueCountNoCase() Function

ListValueCountNoCase() works just the way ListValueCount() does, except that the former ignore capitalization. As with ListValueCount(), substrings are not matched.

```
<cfset myList = "AceSpades,KingSpades,QueenSpades,JackSpades">
<cfset spadeCount = ListValueCount(myList,"Spades")>
<cfset kingSpadeCount = ListValueCountNoCase(myList,"KingSpades")>
<cfset smallKingCount = ListValueCountNoCase(myList,"kingspades")>
<!--- spadeCount is 0, kingSpadeCount is 1, and smallKingCount is 1. --->
<cfdump var="#spadeCount#"><br>
<cfdump var="#kingSpadeCount#"><br>
<cfdump var="#smallKingCount#">
```

ReplaceList() Function

ReplaceList() provides a means to use lists as the sources for a sort of multi-stage replacement. It takes three arguments (and does not take a delimiters argument). The arguments are (1) the string on which to perform the replacement, (2) the list of terms to replace, and (3) the list of terms to use as replacements. If the first element of the first list is found in the string, it is replaced by the first element of the second list, and so on throughout the lists. Note that in the example, the string we're searching for replacements has the characteristics of a list (embedded commas), but in this case it's treated as a simple string. It's the other two variables that are the critical lists in this case.

```
<cfset myString = "AceSpades,KingSpades,QueenSpades,JackSpades">
<cfset lstReplaceThese = "Spades,Ace">
<cfset lstWithThese = "Hearts,Deuce">
<cfset myString = ReplaceList(myString,lstReplaceThese,lstWithThese)>
<!--- myString is "DeuceHearts,KingHearts,QueenHearts,JackHearts" --->
<cfdump var="#myString#">
```

Because the replacements performed by ReplaceList() are performed in order, it's possible to get cascading replacements as the lists are processed left to right. The following code illustrates this cascading behavior in action.

```
<cfset myString = "AceSpades,KingSpades,QueenSpades,JackSpades">
<cfset lstReplaceThese = "Spades,Hearts">
<cfset lstWithThese = "Hearts,Clubs">
<cfset myString = ReplaceList(myString,lstReplaceThese,lstWithThese)>
<!--- myString is "AceClubs,KingClubs,QueenClubs,JackClubs" --->
<cfdump var="#myString#">
```

The first two elements of the lists cause all occurrences of "Spades" to be replaced with "Hearts". The second two elements of the lists then cause all occurrences of "Hearts" to be replaced with "Clubs". The end result is that "Spades" ultimately gets replaced with "Clubs".

With a firm grasp on the capabilities available through the list functions, we can go on to explore some interesting techniques. Before we get completely away from our discussion of the language, though, we need to look at three other considerations: how ColdFusion treats first and last elements of lists (particularly if they are empty), how ColdFusion treats empty list elements in general, and how ColdFusion handles the use of multiple delimiters.

First and Last, Empty and Multiple

List behavior in ColdFusion is fairly straightforward, though there are some tricks to be aware of when dealing with lists. We'll take a quick look at these before moving on to our examples.

First and Last Members

The first and last elements of a list must have some value. If they are empty, they are ignored. To illustrate, consider this code:

```
<cfset myList = ",Two,Three,Four">
<cfset firstMember = ListFirst(myList)>
<!--- Like it or not, firstMember is "Two", rather than an empty value. --->
<cfdump var="#firstMember#">
```

This is consistent with ColdFusion's overall treatment of empty list elements.

Empty Members

You may be familiar with the idea of holding places in a set of fields by using several delimiters in succession. This is frequently done in the comma-delimited file format, when several numeric fields follow one another. It's quite possible to have a sequence of values that looks like this:

```
"Jeff","Peters",,,,"Manassas","VA"
```

Under the comma-delimited file format, this line represents seven fields, with the central three being null. This is not, however, the way ColdFusion sees things. ColdFusion treats empty elements as nonexistent. Therefore, we get the following results:

```
<cfset myList = '"Jeff","Peters",,,,"Manassas","VA"'>
<cfset myListLength = ListLen(myList)>
<!--- myListLength is 4 --->
<cfdump var="#myListLength#">
```

So any time we want to allow "empty" values in a list, but the number of elements in the list is significant, we must use some sort of placeholder string to tell ColdFusion what we really want. The following list shows the example from above with an arbitrary placeholder (the word "null") inserted.

```
<cfset myList = '"Jeff","Peters",null,null,null,"Manassas","VA"'>
<cfset myListLength = ListLen(myList)>
<!--- myListLength is 7 --->
<cfdump var="#myListLength#">
```

The final convolution of all these tricks has to do with the allowance of multiple delimiters to define a list.

Multiple Delimiters

ColdFusion allows the use of multiple delimiters with any list. This means we can use more than one character as the separator between elements of the list. This has very useful applications, but we have to watch out for how this behavior actually works.

Your first reaction when faced with the idea of multiple delimiters, is to think you can specify a string of two or more characters as the delimiter for a list. For example, you might want to use "," as the delimiter when processing a line from a comma-separated values data file. So you do something like this:

```
<cfset myRecord = '"John","Public","New York","NY"'>
<cfset myDelimiter = '","'>
<cfset myArray = listToArray(myRecord,myDelimiter)>
<cfdump var="#myArray#">
```

When you look at the results, you see that myArray is an array with four values, just as you expected. However, let's change the example just slightly and see what happens.

```
<cfset myRecord = '"John","Public","New York, NY"'>
<cfset myDelimiter = '","'>
<cfset myArray = listToArray(myRecord,myDelimiter)>
<cfdump var="#myArray#">
```

In this case, the line stored in myRecord has three fields (according to CSV rules, anyway). But when we run the code, we see that myArray ends up with the same four values we had in the original example. What happened?

ColdFusion interprets multiple delimiters to be interpreted as "mark the beginning of a new element whenever you hit an occurrence of any one of these characters" instead of "mark the beginning of a new element whenever you hit an occurrence of this string of characters". So when we specified "," as the desired delimiters, we were saying "mark the beginning of a new element whenever you hit a double-quote character or a comma". The second double-quote character is actually redundant as far as CF is concerned.

So when the list in the second example above is converted into an array, CF picks up on the comma between New York and NY and separates the two values. There is no difference in the rest of the list because the adjacent delimiters (double-quote and comma) represent empty elements and are ignored by CF.

If we can't specify a multi-character string to use as a delimiter for list functions, what good is having the ability to specify multiple delimiters at all? There are some good uses, and we'll see them in many of the examples in the next section. In fact, the first example, Phone Number Parsing, discusses multiple delimiters.

List Examples

This section provides ten examples of using lists for everyday tasks. Each example uses one or more of the list functions to accomplish its objective(s). Some of the example pages offer more than one solution to each problem.

These examples are not intended to provide the only solution to the problem at hand. Rather, they are here just to start the thoughts flowing about how lists might be used to make your programs more efficient, easier to work with, and easier to maintain.

Example 1: Phone Number Parsing

Phone numbers can be a real challenge sometimes. Let's use U.S. phone numbers as an example. They're not too bad if you happen to get the data collected as a separate Area Code, exchange, and number. But just put a field called "phone" on a form and see how many different formats come back! The following are all commonly in use:

```
(202) 555-1212
202-555-1212
202/555-1212
202.555.1212
```

Parsing these values into area code, exchange, and number can be quite straightforward if we start thinking in terms of lists.

A phone number is really a list of two or three elements. Realizing this, we could extract the area code by doing something like this:

```
<cfset areaCode = Right(ListFirst(myPhone,")"),3)>
```

This would work fine as long as all the phone numbers are in the form of the first instance above. But it's no good if any of the other forms are used. We need a general purpose phone numer parsing strategy:

```
<cfset myDelims = "() -/.">
<cfset areaCode = ListFirst(myPhone,myDelims)>
<cfset exchange = ListGetAt(myPhone,2,myDelims)>
<cfset phoneLine = ListLast(myPhone,myDelims)>
```

By asking the list functions to check for all allowable delimiters (parens, space, hyphen, slash, and dot), we solve the problem of finding out which piece is what.

Example 2:Name Parsing

Very few string parsing problems get more thorny than name parsing. Fortunately, when we're designing applications, we can control name input so we get a given name, a surname, any middle name (or initial), any suffix (like Jr., Sr., III), and an honorific (Mr., Ms., Miss, Col., Gen., etc.) in separate fields.

However, sometimes we have to deal with names as they come in from other sources. In these cases, we need a way to deal with names. In ColdFusion, we can treat a name as a list that uses the space character as its delimiter.

To keep it simple, let's assume a couple of ground rules:

A name may or may not have a middle name or initial

Honorifics and suffixes are contained in separate fields from the name.

The following code uses list functions to process any name that conforms to these rules. (Note: this example uses Hal Helms' CF_QUERYSIM tag to create recordset. For more information on CF_QUERYSIM, see Appendix I. This example also uses a structure. Structures are discussed at length in Chapter 4.)

```
<cf_querysim>
qryMembers
Honorific,MemberName,Suffix
Ms.|Jane Q. Public|null
Mr.|John Doe|III
Mr.|John Quincy Adams|null
Miss|Arlene Zamboni|null
</cf_querysim>

<cfloop query="qryMembers">
  <cfset thisMember = StructNew()>
  <cfset thisMember.firstName = ListFirst(MemberName," ")>
  <cfset restOfName = ListRest(MemberName," ")>
  <cfif ListLen(restOfName," ") GT 1>
    <cfset thisMember.middleName = ListFirst(restOfName," ")>
    <cfset restOfName = ListRest(restOfName," ")>
  </cfif>
  <cfset thisMember.lastName = restOfName>
  <cfdump var="#thisMember#"><br>
</cfloop>
```

Example 3:Decimal Parsing

Dealing with decimal numbers is typically the purview of mathematical functions. Depending on what you're doing, it can be simpler to use a list function to deal with parts of a decimal number. Here's an easy way to return the integer and decimal parts of a number:

```
<cfset myNumber = 3.1415>
<cfset myIntegerPart = ListFirst(myNumber,".")>
<cfset myDecimalPart = ".#ListLast(myNumber,".")#">
<cfdump var="#myIntegerPart#"><br>
<cfdump var="#myDecimalPart#">
```

The odd part about treating decimal numbers as lists is that there is no performance-related reason to avoid this technique. Generally speaking, math functions tend to be faster than string functions, but this is not the case for this example. Using the ListFirst() function is actually faster than using the Int() function. Finding the decimal portion of a number mathematically is comparatively complex with respect to using ListLast():

```
<cfset myDecimalPart = myNumber - Int(myNumber)>
```

Keep in mind, though, that the reason this technique works well is that ColdFusion is a "loosely typed" language. That is, it doesn't care much about whether variables are numbers or strings; CF handles changing the types as necessary. Other languages will not respond well to this sort of treatment.

Example 4:Parse a Text File

Let's revisit Example 2 for a moment, with a small twist in the scenario. Now we'll assume the names we need to import are in a text file given to us by our client. Each line of the file contains one name.

Parsing text files can be an annoying process, but this does not need to be the case. If we recognize the fact that our text file is essentially a list delimited by newline characters, processing it becomes fairly simple:

```
<cffile action="READ" variable="lstFileText"
file="c:\itemp\memberNames.txt">

<cfloop list="#lstFileText#" index="MemberName"
delimiters="#Chr(10)##Chr(13)#">
  <cfset thisMember = StructNew()>
  <cfset thisMember.firstName = ListFirst(MemberName," ")>
  <cfset restOfName = ListRest(MemberName," ")>
  <cfif ListLen(restOfName," ") GT 1>
    <cfset thisMember.middleName = ListFirst(restOfName," ")>
    <cfset restOfName = ListRest(restOfName," ")>
  </cfif>
  <cfset thisMember.lastName = restOfName>
  <cfdump var="#thisMember#"><br>
</cfloop>
```

This is also a nice example of two different list treatments in the same solution. We treat the file as a list delimited by newline characters, and each line as a list delimited by spaces.

Example 5:Parse an E-Mail Address

E-mail addresses can provide a good deal of information about their owners. This example assumes we have a site that needs to do some statistical analysis of the elements within its newsletter mailing list. In order to do this, we need to know the addressee's domain name and whether or not they belong to a commercial domain.

To satisfy these requirements, we need to recognize the e-mail address as two different types of list: one delimited with an @ symbol, and one delimited with dots.

The first requirement, finding the addressee's domain name, uses the first list type:

```
<cfset myEmail = "jeff@grokfusebox.com">
<cfset myDomain = ListLast(myEmail,"@")>
```

The second requirement, deciding whether the domain is commercial, uses the second type. There are two types of commercial domain in our mail list: those that use .com as their top-level domain, and those that are .co domains within their country's top-level domain (john.smith@mycompany.co.uk would be an example of this kind of address). So we need a little conditional logic to figure things out:

```
<cfif ListLast(myDomain,".") EQ "com">
  <cfset isCommercial = True>
<cfelseif ListGetAt(myDomain,ListLen(myDomain,".")-1,".") EQ "co">
  <cfset isCommercial = True>
<cfelse>
  <cfset isCommercial = False>
</cfif>
<cfdump var="#isCommercial#">
```

Example 6:Stacks (Push and Pop)

Stacks are often overlooked in all but the most technical of programming circles. This is unfortunate, as stacks can be very useful things. A stack is essentially a last-in-first-out (LIFO) group of items. The best analogy I've heard is one of those plate holders they have at some buffet-style restaurants (or your college cafeteria). You have to push plates down from the top to put them in the stack, and you can only take them out off the top. That's exactly the way programmatic stacks work. Standard terminology for stacks says that to add an item to the stack, we **push** it onto the stack, and to remove an item from the stack, we **pop** it off.

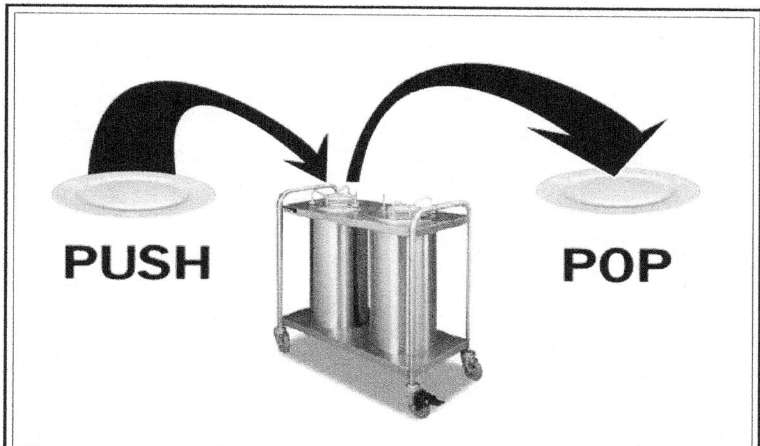

Illustration 2- A Physical Stack

What might we use a stack for? A "breadcrumb" trail is a good use—when following a trail, you can only do one of two things: move forward from where you are (push a new location onto the breadcrumb stack), or go back the way you came (pop the current location off the breadcrumb stack). List functions can help us use stacks in ColdFusion.

For our approach, we'll make the left end of the list act as the stack's interface. That is, when we push an item onto the stack, it goes on the left end, shifting everything else to the right. When we pop an item off the stack, the item on the left end is the one we get, shifting the remaining items to the left.

The ListPrepend() function is tailor-made to act as our push behavior:

```
<cfset myStack = "">
<cfset myStack = ListPrepend(myStack,"Foyer")>
<cfdump var="#myStack#">
```

The pop behavior is slightly more complex, because we need to read the "top" item of the stack and remove it from the stack as well:

```
<cfset lastRoom = ListFirst(myStack)>
<cfset myStack = ListRest(myStack)>
<cfdump var="#lastRoom#"><br>
<cfdump var="#myStack#">
```

This would be a good place to introduce a couple of user-defined functions (UDFs) to do our push and pop behaviors:

```
<cfscript>
  function Push(theList,theValue){
    variables[theList] = ListPrepend(variables[theList],theValue);
  }
  function Pop(theList){
    popValue = ListFirst(variables[theList]);
    variables[theList] = ListRest(variables[theList]);
    return popValue;
  }
</cfscript>

<cfset variables.myStack = "foyer">
<cfscript>
  Push("myStack","hall");
  Push("myStack","dining room");
  lastItem = Pop("myStack");
</cfscript>

<cfdump var="#lastItem#"><br>
<cfdump var="#myStack#">
```

Using this technique, we could build a sequence of pages to complete an action (sometimes referred to as a "wizard") with Next and Previous on each page. The Next button would push an item onto the breadcrumb stack as part of its processing, and the Previous button would pop an item off the stack.

Example 7:Using CONTAINS

In the section on ColdFusion list functions, I detailed the workings of the ListContains(), ListContainsNoCase(), ListFind() and ListFindNoCase() functions.

You may also be aware of ColdFusion's CONTAINS string comparison operator. This operator is an easy way to check if one string has a specific substring within it:

```
<cfset myString = "the quick brown fox">
<cfset stringToFind = "ick">
<cfif myString CONTAINS stringToFind>
  I found the stringToFind.
</cfif>
```

Where the four functions mentioned above perform matches within the context of the list, CONTAINS completely ignores whether the string to be searched is a list and treats it as a simple string.

This means we can check for substrings that cross more than one element of the list. For example, let's stick with the sequence of rooms we used in Example 6. If we have a requirement that everyone move, at some point in their tour of our house, from the hall to the dining room to the kitchen, we could do something like this:

```
<cfif myStack CONTAINS "hall,dining room,kitchen">
  You've satisfied the tour requirements
</cfif>
```

Be careful when using CONTAINS, though, as unexpected results are very easy to encounter. Consider the previous code sample. It would test true if the value of myStack was "west hall,dining room,kitchen pantry". This is not what we wanted. So beware: CONTAINS can be very useful, but use care.

Example 8:Build a Math Sequence

Calculators and instructional applications often need to build sequences of values and operators. These expressions can be constructed using list functions and the ability to specify the delimiter. We can build a sequence by using the desired operator as the delimiter each time we add a value to the list. After we're done building the expression, we can output both the expression and its solution (using the Evaluate() function).

```
<cfset myEquation = "">
<cfscript>
  myValue = 5;
  myOperator = "+";
  myEquation = ListAppend(myEquation,myValue,myOperator);

  myValue = 4;
  myOperator = "-";
  myEquation = ListAppend(myEquation,myValue,myOperator);

  myValue = 3;
  myOperator = "+";
  myEquation = ListAppend(myEquation,myValue,myOperator);
</cfscript>

<cfdump var="#myEquation#"><br>
<cfoutput>
#myEquation# = #Evaluate(myEquation)#
</cfoutput>
```

Example 9:Count Words

Counting words in a piece of text prior to the advent of word processing was not an easy task, thanks to the extreme variation in the length of words. Consequently, methods other than actual word count were developed. The most common of these was the average word length method. This method stated that the average word was 5 characters long. Since each line of a piece of text had a maximum of 80 characters, a decent estimate of the word count could be calculated by counting lines and multiplying by 16.

Fortunately for accuracy, we can easily count actual words in a piece of text. By now you've probably guessed that I see this as a simple case of interpreting the paragraph as a space-delimited list. This is partially true; in addition to spaces, we need to account for any whitespace characters when assigning delimiters:

```
<cfset myDelimiters = " #Chr(10)##Chr(13)##Chr(9)#">
<cfset wordCount = ListLen(myParagraph,myDelimiters)>
<cfdump var="#wordCount#">
```

The myDelimiters variable is set to include the space, line feed [Chr(10)], carriage return [Chr(13)], and tab [Chr(9)] characters, then the ListLen() function is used to return an accurate word count.

> **Note:** DOS/Windows-format text files use both carriage return and line feed characters at the end of each line of text. Unix-format files (also used on Mac OS X) use only the line feed character, referred to as the "newline" character.

Example 10:Loop over a list

In Example 4, I used <cfloop> to loop over each element of a list, in that case each line of a text file. I wanted to use one more example to talk specifically about looping over lists.

When looping over a list, we use the **list**, **item**, and (optionally) **delimiters** attributes. The list attribute specifies a list. Usually the list has been constructed before the <cfloop> call, so we reference it by name:

```
list="#myList#"
```

The item attribute lets us assign a name we'll use to represent the element of the list for each iteration of the loop. This can be a plain string or a reference to a variable. It is typical to use a variable name such as "thisMember" to help make the code within the loop easier to read.

```
Item="thisMember"
```

Finally, of course, we can use the delimiters attribute if the list uses delimiters other than the comma.

Here's a quick example that shows a loop over a credit card numer to ensure that each segment has four digits:

```
<cfset ccNumber = "1234 5678 9012 3456">
<cfset badNumber = False>
<cfloop list="#ccNumber#" index="thisSegment" delimiters=" ">
<cfif Len(thisSegment) NEQ 4>
  badNumber = True
</cfif>
</cfloop>
<cfdump var="#ccNumber#"><br>
<cfdump var="#badNumber#">
```

Last on the List

Lists are one of the very useful complex data types available to you as a ColdFusion programmer. Getting familiar with list handling will open many doors to programming solutions that are not possible without using lists.

A list is simply a string that uses a delimiter to separate elements. The default delimiter is the comma, but you can specify any delimiter you wish. ColdFusion has a strong set of functions to allow easy manipulation of lists. Remember, though, that lists are just strings, and you don't necessarily need to use a list function to work on them. It is sometimes quicker to do a string concatenation:

```
<cfset myList = "#myList#, newValue">
```

than it is to do a list function:

```
<cfset myList = ListAppend(myList, "newValue")>
```

Whether you use a list function or a simple string operation is up to you as a coder; the fact that lists are just specialized strings opens the door to any of a wide variety of manipulation techniques.

Lists are very powerful, but are not the solution to every problem where we need more than simple datatypes. The next chapter examines arrays, which I treat as something like high-powered lists. Arrays are fast, flexible, and highly useful in looping situations. If there was benefit in learning about lists in this chapter, there's gold to be mined from Chapter 3 – Arrays.

1. Arrays

What Is An Array?

An array is a complex data type that lets you store more than one value. Most programming languages have the capability to process arrays. ColdFusion's support for arrays includes a large collection of array processing functions.

To get a good handle on arrays, think about an array as a set of containers. Each container can hold a value. The array is referred to by name, and each of its containers (or elements, as they are called) can be referred to by a positional number or index. A good model of an array in real life is a six-pack of beverages. A six-pack is an array with six elements (cans); each element holds a value (some amount of beverage).

In the world of ColdFusion, the content of an element can be a simple or complex value. That is, you can store a string, number, or Boolean value in an array element, but you can also store a list, a structure or even another array in an array element. We'll get to these more complex arrangements a little further on.

Regardless of what value is stored in an array element, it is referenced in the same way: by its index number. Let's go back to our six-pack example for a moment. Imagine a very unusual six-pack that has the following cans of beverage in it:

Cola
Citrus soda
Beer
Water
Grape soda
Orange juice

If I want to ask for the can of grape soda, an array won't let me say, "I'd like the can of grape soda, please". Instead, I must say, "Give me the contents of the fifth can, please." In ColdFusion code, setting up the array and asking for the can of grape soda would look like this:

Tag Method:

```
<cfset arySixPack = ArrayNew(1)>
<cfset ArrayAppend(arySixPack, "Cola")>
<cfset ArrayAppend(arySixPack, "Citrus soda")>
<cfset ArrayAppend(arySixPack, "Beer")>
<cfset ArrayAppend(arySixPack, "Water")>
<cfset ArrayAppend(arySixPack, "Grape Soda")>
<cfset ArrayAppend(arySixPack, "Orange juice")>
<cfoutput>#arySixPack[5]#</cfoutput>
```

CFScript Method:

```
<cfscript>
 arySixPack = ArrayNew(1);
 ArrayAppend(arySixPack, "Cola");
 ArrayAppend(arySixPack, "Citrus soda");
 ArrayAppend(arySixPack, "Beer");
 ArrayAppend(arySixPack, "Water");
 ArrayAppend(arySixPack, "Grape Soda");
 ArrayAppend(arySixPack, "Orange juice");
</cfscript>
<cfoutput>#arySixPack[5]#</cfoutput>
```

ColdFusion 8 Method:

```
<cfscript>
 arySixPack = ["Cola",
             "Citrus soda",
             "Beer",
             "Water",
             "Grape Soda",
             "Orange juice"];
</cfscript>
<cfoutput>#arySixPack[5]#</cfoutput>
```

So the "can" of grape soda, which is in the fifth element of the array, can be referenced by using the expression arySixPack[5].

In fact, all array elements are referenced in this fashion: the name of the array, followed by the element's index number enclosed in square brackets.

The value of the index number can also be contained in a variable or an expression. So the following code produces the same result:

```
<cfset whichElement = 5>
<cfoutput>#arySixPack[whichElement]#</cfoutput>
```

We'll be making thorough use of expressions to reference array elements in the examples later in this chapter. Before we get there, let's take a closer look at the characteristics and behavior of arrays. We'll start with one-dimensional arrays and move on to multi-dimensional arrays.

One-Dimensional Arrays

Arrays can be defined in one or more dimensions. To get started, we'll look at one-dimensional arrays. Later, when we look at multi-dimensional arrays, the characteristics and behaviors we see in one-dimensional arrays will carry through.

A one-dimensional array is somewhat like a list in that it is a set of several values stored in a single variable. But this is where the similarity between lists and arrays ends. Unlike a list, an array is not a simple string variable. Arrays are, by their nature, complex data; we can't use simple datatype manipulation on an array.

Because array elements are referenced by their index numbers, it is possible to create an array that has "holes", or empty elements. The following code creates an array that has a hole at index 4:

```
<cfset aryFriends = ArrayNew(1)>
<cfscript>
  aryFriends[1] = "Alan";
  aryFriends[2] = "Brian";
  aryFriends[3] = "Chip";
  aryFriends[5] = "Evan";
</cfscript>
```

If we try to use aryFriends[4] (within a <cfoutput> block, for example), ColdFusion lets us know there's a problem with this error message (CFMX message shown; CF5 and prior have a different message):

```
Element 4 is undefined in a Java object of type class coldfusion.runtime.Array
```

This potential for holes becomes important when we attempt to loop over an array. We'll work with looping a little later; first we'll go over the basics of array handling in one dimension.

Array How-To (One Dimension)

Working with arrays is fairly simple business once you're familiar with the basics. We'll go through creating an array, adding and deleting elements, and sorting an array. Then we'll look at looping over an array to perform operations on each element. Finally, we'll need a way to destroy an array once we're done with it.

Create an Array

Creating an array in ColdFusion is very simple—we just use the ArrayNew() function. ArrayNew() takes the number of dimensions for the array as its only argument.

To create a one-dimensional array, we can use this syntax:

```
<cfset aryMenuItems = ArrayNew(1)>
```

We can also create an array inside a <cfscript> block:

```
<cfscript>
 aryMenuItems = ArrayNew(1);
</cfscript>
```

As of ColdFusion 8, the ArrayNew() function is no longer necessary. As shown in the *What Is an Array?* section, arrays can be created with the new direct syntax which, typically, creates and populates an array in one statement. However, we can also use the new syntax to create an empty array, so:

```
<cfset aryMenuItems = []>
```

However, if there is a need to create an empty array, I prefer the ArrayNew(1) syntax, because it is self-documenting. The newer syntax requires someone reading the code to know that square brackets represent an array.

Speaking of self-documenting code, the format of the array variable name is a convention I like to use. The three-letter "ary" prefix indicates that the variable is an array. This is not a requirement of ColdFusion. It just makes the variable easier to recognize as an array when I revisit the code. Some people prefer to use a suffix, for example menuItemsArray. Either way is fine.

Add Elements To an Array

We can add elements to an array in several ways. An element's value can be set directly, using its index number:

```
<cfset aryMenuItems[1] = "File">
```

or

```
<cfscript>
 aryMenuItems[1] = "File";
</cfscript>
```

> **NOTE:** The direct syntax introduced in ColdFusion 8 makes the array creation and population a single statement:
> <cfset aryMenuItems = [File]>

The index number method is workable, but only in cases where we know what index number we want to use. When first populating an array, it would be much easier if we had a way to say "add an element to the end of this array", in the same way we appended elements to a list with the ListAppend() function.

Fortunately, the designers of CFML recognized this fact as well, and gave us the ArrayAppend() function. The ArrayAppend() function takes as its arguments the name of the array and the value to append as a new element, so:

```
cfset  ArrayAppend(aryMenuItems,"File")>
```

or

```
<cfscript>
 ArrayAppend(aryMenuItems,"File");
</cfscript>
```

With ArrayAppend(), we don't need to know anything about the index numbers that already exist in the array, so we can easily populate an array with various values.

Sometimes we need to create an array of a specific size and populate it with the same value in every element. This can be done very easily with the ArraySet() function.

As an example, let's say we need to work with values that represent the results of six tests. If the test was not successfully completed, its result should be zero. We can set up an array of six elements with zero as the default value like this:

```
<cfset aryTests = ArrayNew(1)>
<cfset ArraySet(aryTests,1,6,0)>
```

or

```
<cfscript>
 aryTests = ArrayNew(1);
 ArraySet(aryTests,1,6,0);
</cfscript>
```

The ArraySet() function takes as its arguments the name of the array to be populated, the starting index, the ending index and the value to insert into each element.

In ColdFusion 8, we could also use the following syntax to get the same result:

```
<cfscript>
 aryTests = [0,0,0,0,0,0];
</cfscript>
```

Once we have an array populated, there may be times when we want to delete elements from it. The next section covers deleting elements from an array.

Delete Elements From an Array

Keeping in mind that elements of an array are referenced by the index numbers, deleting them is an easy task. If we want to delete the second element in aryMenuItems, we just do:

```
<cfset ArrayDeleteAt(aryMenuItems,2)>
```

or

```
<cfscript>
 ArrayDeleteAt(aryMenuItems,2);
</cfscript>
```

Some programmers "delete" elements from an array by setting their values to an empty string like this:

```
<cfset aryMenuItems[2] = "">
```

This technique does not actually delete the element, because an empty string is a value. It does, however, prevent reindexing of the array. This may be a desirable behavior depending on what you're doing with your array. If you delete an array element with ArrayDeleteAt(), then try to reference that element, you'll get the value of the element that followed that element before the deletion. If you use the "set to empty string" method and then try to reference the element, you'll get an empty string returned as the value of the element. So if the index of the element is meaningful, it's better to use the empty string method. On the other hand, if the index is not meaningful, it's better to use the ArrayDeleteAt() method.

Sometimes we want to delete all the elements from an array. This is easily done with the ArrayClear() function, which clears (deletes) all the elements in an array:

```
<cfset ArrayClear(aryMenuItems)>
```

or

```
<cfscript>
 ArrayClear(aryMenuItems);
</cfscript>
```

Now that we know how to add and delete elements from an array, we can look at some more useful array manipulations. We'll start with sorting.

Sort an Array

Sorting is one of the most common operations performed on collections of values. There have been volumes written on algorithms to perform sorts. ColdFusion lets us sort without having to worry about how the guts work. All we need to know is the ArraySort() function:

```
<cfset ArraySort(aryMenuItems,"textnocase")>
```

or

```
<cfscript>
 ArraySort(aryMenuItems,"textnocase")
</cfscript>
```

More details on the ArraySort() function can be found in the Array Functions section later in this chapter.

Loop over an array

Looping is one of the most powerful means for employing arrays. Looping simply means performing a set of operations on each member of a group—in this case, each element of an array.

To loop over an array in ColdFusion, we use the <cfloop> tag. <cfloop> can take a variety of arguments, depending on what type of construct we need to loop over. In the case of an array, we need to use the from, to, and index attributes.

The from attribute lets us specify the index number to start from; the to attribute lets us specify the last index number we want to use. The index attribute lets us specify a variable name that will reference the current index number within the loop. It is common practice to use the letter i as the variable name for a loop's index.

To loop over an array, we typically start with an index of 1, and we want to continue looping through the end of the array. We do this by using the ArrayLen() function to return the number of elements in the array over which we're looping.

Putting these attributes together, we can loop over the aryMenuItems array with the following syntax:

```
<cfloop from="1"
       to="#ArrayLen(aryMenuItems)#"
     index="i">
 <cfoutput>#aryMenuItems[i]#</cfoutput>
</cfloop>
```

This would display the contents of each element of the array.

We can also perform loops within <cfscript> blocks, using the for looping construct (note that we can't use <cfoutput> from within a <cfscript> block, so we're adding values to a list for output outside the script block):

```
<cfscript>
 lstTitles = "";
 for (i = 1;
     i LTE ArrayLen(aryMenuItems);
     i = i +1){
   lstTitles = ListAppend(lstTitles,aryMenuItems[i]);
   }
</cfscript>
<cfoutput>#lstTitles#</cfoutput>
```

The three parts of the for-loop specification (the parts inside the parentheses) specify a starting variable (i=1), a condition for which the loop will process as long as the condition is true (i LTE ArrayLen(aryMenuItems)), and an operation to perform at the end of every iteration of the loop (i = i +1). The example above demonstrates a very simple looping procedure, but loops can be much more powerful, allowing us to perform any number of operations within the body of the loop.

There are also other types of loops available in <cfscript>. In addition to the for loop, we can also perform while and do-while loops. A while loop continues to process as long as a specified condition is true; the condition is evaluated at the beginning of the loop. A do-while loop continues to process as long as a specified condition is true; the condition is evaluated at the end of the loop. Examples of these two techniques are shown below.

```
<cfscript>
 a = ArrayNew(1);
 i = 11;
 while (i LT 10)
 {
  a[i] = i;
  i = i + 1;
 }
</cfscript>

<cfdump var="#a#">
```

and the do-while version:

```
<cfscript>
 a = ArrayNew(1);
 i = 11;
 do
 {
  a[i] = i + 5;
  i = i + 2;
 }
 while (i LT 10);
</cfscript>

<cfdump var="#a#">
```

If you run the two example above, you'll notice that in the first case, where the test happens at the beginning of the loop, the array named a remains empty, because the value of i is greater than 10. In the second case, where the test happens at the end of the loop, a gets one element defined. Even though the value of i is greater than 10, it doesn't get tested until after the first element is defined.

Well, we've now been through a fairly thorough treatment of working with one-dimensional arrays. The only thing we have left to do is to get rid of them once we're done with them.

Destroy an Array

ColdFusion doesn't really provide a means to destroy variables (with the exception of client variables, but that's another issue). As mentioned above, we can delete all the elements in an array by using the ArrayClear() function. Some programmers "destroy" arrays by saving an empty string to the name of the array. This effectively transforms the array into a string variable with an empty string as its value:

```
<cfset aryMenuItems = "">
```

or

```
<cfscript>
 aryMenuItems = "";
</cfscript>
```

If we now test the variable aryMenuItems with the IsArray() function, the result we'll get is NO.

Multiple Dimensions

Arrays in two or more dimensions are not much more complex than arrays in one dimension. The trick to the whole game is to recognize that a multi-dimensional array is simply a nested set of arrays. To clarify, a two-dimensional array is an array whose elements are themselves arrays.

The nature of multi-dimensional arrays might become a bit clearer if we examine how arrays are created. The ArrayNew() function takes one argument, which is an integer from 1 to 3. This argument specifies the number of dimensions to be used for the new array. So, to define a two-dimensional array, the typical syntax looks like this:

```
<cfset aryTeammates = ArrayNew(2)>
```

or

```
<cfscript>
 aryTeammates = ArrayNew(2);
</cfscript>
```

Let's say we need to store three sets of data in our new array (three values in the first dimension). If we knew this in advance, we could define the array this way:

```
<cfset aryTeammates = ArrayNew(1)>
<cfloop from="1" to="3" index="i">
 <cfset aryTeammates[i]= ArrayNew(1)>
 <cfset aryTeammates[i][1] = i>
 <cfset aryTeammates[i][2] = "A">
</cfloop>
```

In ColdFusion 8, we could use a more direct syntax to do the same thing.

```
<cfset aryTeammates = ArrayNew(1)>
<cfloop from="1" to="3" index="i">
 <cfset aryTeammates[i] = [i,"A"]>
</cfloop>
```

The end result is a two-dimensional array that has three elements in the first dimension.

Why is it important to understand the nested nature of mult-dimensional arrays? The ArrayNew() function limits array creation to three dimensions. If we know that a multi-dimensional array is just a nested group of arrays, we can create arrays in as many dimensions as we want.

In turn, if we have the ability to manipulate arrays in any number of dimensions without the restrictions of the ArrayNew() function, we've removed one of the arbitrary limitations of CFML. The next sections are essentially duplicates of the preceding Array How-To, with examples for multi-dimensional arrays.

Array How-To (Multiple Dimensions)

The key to dealing with multi-dimensional arrays is very simple: a multi-dimensional array is just a set of nested one-dimensional arrays, no matter how many dimensions it has.

To put it another way, the contents of every element of a multi-dimensional array is another array, with the exception of the last (rightmost) dimension. The following illustration shows a two-dimensional array that stores values for an x,y coordinate system. Note that only the elements in the second dimension contain numbers; the elements of the first dimension contain arrays.

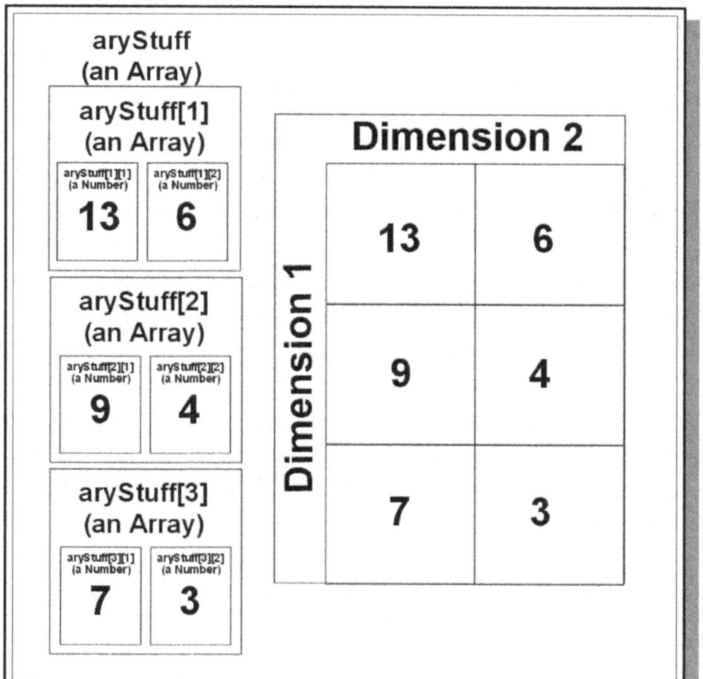

Illustration 3: Two Views of a Two-Dimensional Array

This is the case regardless of how many dimensions the array has—every element in each dimension is an array, except for the last dimension. Knowing this, we can apply any of ColdFusion's array functions to any element in a multi-dimensional array, with the exception of the elements in the last dimension. So, in the case of the example in Illustration 1, the following code is a valid way to add a fourth set of coordinates:

```
<cfset aryStuff[4] = ArrayNew(1)>
<cfset aryStuff[4][1] = 17>
<cfset aryStuff[4][2] = 8>
```

Or, with ColdFusion 8:

```
<cfset aryStuff[4] = [17,8]>
```

With all that said, we can move on to look at specifics for creating, adding and deleting elements, and looping with multi-dimensional arrays. Then we'll wrap up this chapter with some examples of arrays in action. We'll start with a look at a speedy method for creating multi-dimensional arrays.

Create an Array (Multiple Dimensions)

Keeping in mind that multi-dimensional arrays are simply arrays of arrays, the creation process is just the one-dimensional process repeated for as many dimensions as we need. The process can be implemented very nicely with the use of loops.

As an example, let's say we need to create an array to store the temperature for the first 20 days of each month for the last 10 years. This means we need a three-dimensional array (days by months by years) to manipulate the data.

We can create this multi-dimensioned array very quickly with a looping technique. Since we know that we're going to have 10 years, 12 months in each year, and 20 days in each month, the array can be created with the following code (we'll default each value to 0 to make sure all our numeric manipulations work):

```
<cfset aryTemperatures = ArrayNew(1)>
<cfloop from="1" to="10" index="thisYear">
  <cfset aryTemperatures[thisYear] = ArrayNew(1)>
  <cfloop from="1" to="12" index="thisMonth">
    <cfset aryTemperatures[thisYear][thisMonth] = ArrayNew(1)>
      <cfloop from="1" to="20" index="thisDay">
        <cfset aryTemperatures[thisYear][thisMonth][thisDay] = 0>
      </cfloop>
  </cfloop>
</cfloop>
```

Of course we could also have created this specific array with the syntax `<cfset aryTemperatures = ArrayNew(3)>` and populated it with zeros through nested loops. Though true for this example, the idea is to see that we can create any necessary number of dimensions through the use of loops; we're not constrained to the limitation of 3 dimensions imposed by the ArrayNew() function.

We could also create this array with hard-coding in the direct syntax of ColdFusion 8, but it would become very cumbersome, as we'd have to specify 10x12x20 = 2400 elements. In this particular case, the syntax above is fine in CF8 as well.

Add Elements To an Array (Multiple Dimensions)

Adding elements to a multi-dimensional array is only slightly more complex than adding elements to a one-dimensional array. The thing to keep in mind is that we only store actual data in the last dimension. Adding an element to any dimension other than the last one calls for adding an array.

So, to add an element to our aryTemperatures array (let's say we want to store a temperature of 34 degrees for the twenty-first day of January of the first year), we could just do this:

```
<cfset aryTemperatures[1][1][21] = 34>
```

Depending on what we intend to do with our data, though, this might cause a problem. We now have 21 elements in January of the first year, and only 20 elements in each of the other months. Since we're setting default values to zero, we should create a 21st day for each month, and set its value to zero, then set the value of the first January 1st to 34:

```
<cfloop from="1" to="10" index="thisYear">
 <cfset aryTemperatures[thisYear] = ArrayNew(1)>
 <cfloop from="1" to="12" index="thisMonth">
  <cfset aryTemperatures[thisYear][thisMonth][21] = 0>
 </cfloop>
</cfloop>
<cfset aryTemperatures[1][1][21] = 34>
```

So adding elements to a multi-dimensional array isn't too tough, as long as we keep in mind the idea that we're operating in a system of dimensions. Deleting elements, on the other hand, can get to be a bit tricky.

Delete Elements From an Array (Multiple Dimensions)

As noted in the Delete Elements From an Array section above, when we use the ArrayDeleteAt() function to delete an element of an array in ColdFusion, the array closes up around the deleted element. That is, each element to the right of the deleted element shifts one step to the left, acquiring a new index number that is one less than its index number prior to the deletion.

This behavior makes sense from the point of view that arrays are designed to enable looping behaviors in code; we don't want holes all over the place to cause problems when we're running loops. However, it does mean that we need to keep the behavior in mind when we start deleting elements from multi-dimensional arrays.

To illustrate the potential pitfalls with deleting elements of multi-dimensional arrays, we'll use a smaller example than our temperatures array, so we can display it easily. This array stores tic-tac-toe games played over time:

```
<cfset aryGames = ArrayNew(3)>
<cfset aryGames[1][1][1] = "X">
<cfset aryGames[1][1][2] = "O">
<cfset aryGames[1][1][3] = "X">
<cfset aryGames[1][2][1] = "O">
<cfset aryGames[1][2][2] = "X">
<cfset aryGames[1][2][3] = "O">
<cfset aryGames[1][3][1] = "X">
<cfset aryGames[1][3][2] = "O">
<cfset aryGames[1][3][3] = "X">
<cfset aryGames[2][1][1] = "O">
<cfset aryGames[2][1][2] = "X">
<cfset aryGames[2][1][3] = "O">
<cfset aryGames[2][2][1] = "X">
<cfset aryGames[2][2][2] = "O">
<cfset aryGames[2][2][3] = "X">
<cfset aryGames[2][3][1] = "O">
<cfset aryGames[2][3][2] = "X">
<cfset aryGames[2][3][3] = "O">
```

The illustration below shows this array divided into the two games it represents, aryGames[1] and aryGames[2].

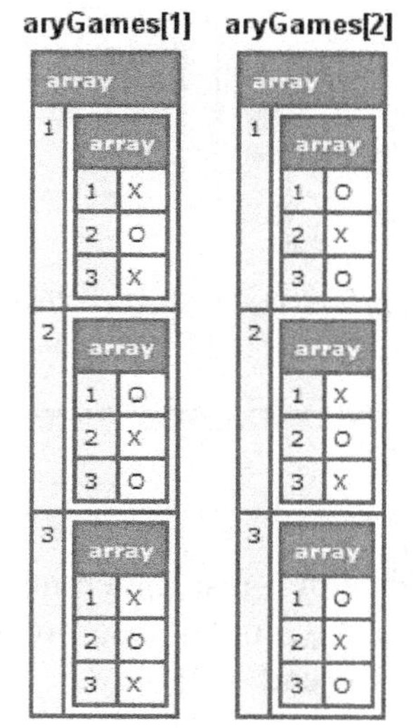

Illustration 4: aryGames Array

Now let's delete an element of the array—specifically the second game's square at row 2, column 2. This is done with this line:

```
<cfset ArrayDeleteAt(aryGames[2][2],2)>
```

When the element is deleted, the next element slides over and closes the gap. So instead of having a missing value at aryGames[2][2][2], we end up with the value from aryGames[2][2][3] in aryGames[2][2][2], and no element at aryGames[2][2][3]. The result is shown below.

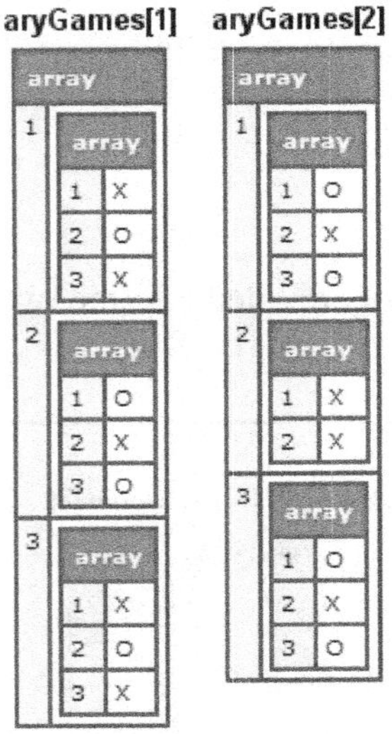

Illustration 5: aryGames After Deletion

Again, this is not necessarily bad behavior—just something to be kept in mind when working with multi-dimensional arrays. Obviously, in this case, we've lost some valuable information: the state of the board at the end of game 2.

Loop Over an Array (Multiple Dimensions)

Looping over multi-dimensional arrays is just a matter of nesting loops, in just the same way that the array itself is just a set of nested one-dimensional arrays. The code in the previous sections has illustrated this style of looping so we won't develop a separate example here. There are a few pointers that will make looping over multi-dimensional array a bit easier, though.

When using nested loops, indent your code to make it more easily readable. Take the following nested loop:

```
<cfloop from="1" to="10" index="i"><cfloop from="1" to="10" index="j"><cfset aryStuff[i][j] =
Rand()></cfloop></cfloop>
```

The code itself is fine; it's just a bit difficult to follow what's going on. On the other hand, the same code with some indentation is more readable:

```
<cfloop from="1" to="10" index="i">
  <cfloop from="1" to="10" index="j">
    <cfset aryStuff[i][j] = Rand()>
  </cfloop>
</cfloop>
```

Now the behavior of the code block is much more clear—there are two levels of looping involved, and the index of each loop is easy to find.

In similar fashion, the use of descriptive index variables will help keep track of multi-dimensional arrays, particularly if you're dealing in more than two dimensions. In the aryTemperature example in the previous section, we used the index variables `thisYear`, `thisMonth`, and `thisDay`. These variables make it much easier to keep track of which loop operates on which dimension than using single-letter variables like `i`, `j`, and `k`: Single-letter index variables are very common practice for loops in many languages, dating back to when memory was much more limited and variable names needed to be terse. There's nothing inherently wrong with them, so if you're accustomed to using them and comfortable doing so, carry on. Otherwise, be descriptive.

ColdFusion Array Functions

As with the previous chapter on lists, this chapter provides a complete list of ColdFusion array functions and a brief look at the behavior of each. We'll look at some of the tips and tricks that aren't necessarily mentioned in the Adobe documentation.

ArrayAppend() Function

The ArrayAppend() function adds an element to the end of an array. You don't need to know how long the array is. This function operates on one-dimensional arrays, but you can add elements to a multi-dimensional array using ArrayAppend(). The key is to remember that multi-dimensional arrays are just nested one-dimensional arrays.

ArrayAppend() takes two arguments: the array to append to, and the value to append. It returns a boolean TRUE if the append is successful. (This means ArrayAppend() always returns TRUE—making us wonder why it returns anything at all. This, however, is a discussion for another book.) Both of the following examples are valid uses of ArrayAppend():

```
<cfset temp = ArrayAppend(aryStudents,"Fred")>
<cfset ArrayAppend(aryGames[1],ArrayNew(1))>
```

Note that the second example is adding a new element to the multi-dimensional array named aryGames. It also does not use a temporary variable to store the return value of ArrayAppend(). This is a common practice among CF developers, recognizing that ArrayAppend() always returns TRUE if it doesn't throw an error.

ArrayIsEmpty() Function

The ArrayIsEmpty() function tests an array for content. If the array has no elements, the function returns true; otherwise it returns false. If an argument of a type other than array is passed to the function, it throws an error:

```
<cfset aryParts = ArrayNew(1)>
<cfset stuff = 0>
<cfoutput>#ArrayIsEmtpy(aryParts)#</cfoutput> <!--- Returns true --->
<cfset ArrayAppend(aryParts,"10115")>
<cfoutput>#ArrayIsEmpty(aryParts)#</cfoutput> <!--- Returns false --->
<cfoutput>#ArrayIsEmpty(stuff)#</cfoutput> <!--- Throws error --->
```

ArrayPrepend() Function

The ArrayPrepend() function adds an element to the beginning of an array. This function operates on one-dimensional arrays, but you can add elements to a multi-dimensional array using ArrayPrepend(). The key is to remember that multi-dimensional arrays are just nested one-dimensional arrays. Beware, though—prepending to a multi-dimensional array can mess up the organization of your data if you're not careful about what you're doing—see the section on multi-dimensional arrays for more information on multi-dimensional positioning.

ArrayPrepend() takes two arguments: the array to prepend to, and the value to prepend. It returns a boolean TRUE if the append is successful. (Like ArrayAppend(), ArrayPrepend() always returns TRUE.) Both of the following examples are valid uses of ArrayPrepend():

```
<cfset temp = ArrayPrepend(aryStudents,"Fred")>
<cfset ArrayPrepend(aryGames[1],ArrayNew(1))>
```

Note that the second example is inserting a new element into the multi-dimensional array named aryGames.

ArraySwap() Function

The ArraySwap() function trades the places of two elements of the specified array. It takes three arguments: the array to use, the index of the first element, and the index of the second element. It returns true if the arguments are all valid, and throws an error otherwise.

In this example, notice that the ArraySwap() line causes the two values in elements 3 and 5 to trade places:

```
<cfset aryStuff = ArrayNew(1)>
<cfscript>
 ArrayAppend(aryStuff,"one");
 ArrayAppend(aryStuff,"two");
 ArrayAppend(aryStuff,"three");
 ArrayAppend(aryStuff,"four");
 ArrayAppend(aryStuff,"five");
 ArraySwap(aryStuff,3,5);
</cfscript>

<cfdump var="#aryStuff#">
```

The output from this example is shown below; you can see that the values for the third and fifth elements have been swapped.

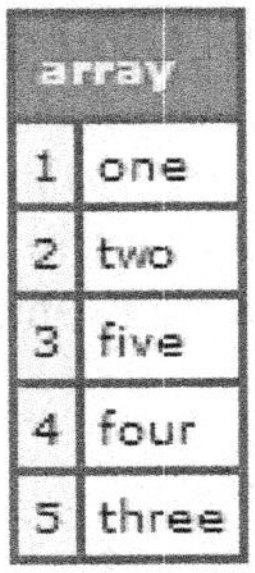

Illustration 6: aryStuff After Swap

ArrayAvg() Function

The ArrayAvg() function returns the average (arithmetic mean) of the elements in the array. The values of all the elements are added up, and the sum is divided by the number of elements.

This function requires that all the elements of the array have numeric values. If there is an empty element in the array, or an element that is not numeric, the function will throw an error. Integers and decimal values are both allowed; if the result is an integer, the function returns an integer, regardless of whether decimal values were present in the array. The following example will return 3, not 3.0 as you might guess due to the presence of a decimal value in the second element:

```
<cfset aryStuff = ArrayNew(1)>
<cfscript>
 ArrayAppend(aryStuff,1);
 ArrayAppend(aryStuff,2.0);
 ArrayAppend(aryStuff,3)
 ArrayAppend(aryStuff,4);
 ArrayAppend(aryStuff,5);
 fltAvg = ArrayAvg(aryStuff);
</cfscript>
<cfdump var="#fltAvg#">
```

ArrayLen() Function

The ArrayLen() function returns the number of elements in an array, regardless of whether they are empty. The following code shows the creation of an array with an empty element in the fourth position. The ArrayLen() function still returns 5.

```
<cfset aryStuff = ArrayNew(1)>
<cfscript>
 ArrayAppend(aryStuff,1);
 ArrayAppend(aryStuff,2);
 ArrayAppend(aryStuff,3);
 aryStuff[5] = 5;
 intLength = ArrayLen(aryStuff);
</cfscript>
<cfdump var="#intLength#">  <!--- Returns 5 --->
```

If you need to loop over an array that might have empty elements, you should use a function like IsNumeric(), IsDate(), or IsSimpleValue() to check for a value before you try to do anything with it.

ArrayResize() Function

The ArrayResize() function appends empty elements to the end of an array to make the array's length match the specified value. The function takes two arguments: the array and an integer to specify the minimum new size. It returns true if the arguments are valid. If the size argument is a value smaller than the number of elements in the array, the array is not changed.

ArrayToList() Function

The ArrayToList() function converts an array to a list, using the specified delimiter for the list. It takes one required argument – the array to convert – and one optional argument – the delimiter to use in the new list. If a variable of type other than array is used for the first argument, an error is thrown. Likewise, if a variable of type other than a simple datatype is used for the second argument, an error is thrown.

One interesting behavior of ArrayToList(): if you use a delimiter with more than one character, the function inserts it between elements of the list in the same manner as a single-character delimiter. This introduces some interesting possibilities. For example, if we have an array of SQL criteria, we can assemble a valid WHERE clause very easily:

```
<cfset aryParms = ArrayNew(1)>
<cfscript>
 ArrayAppend(aryParms,"firstName > 'A'");
 ArrayAppend(aryParms,"age < 55");
 ArrayAppend(aryParms,"gender = 'F'");
</cfscript>
<cfset whereClause = "WHERE #ArrayToList(aryParms,' AND ')#">
<cfdump var="#whereClause#">
```

Though it's not usual, we can also use numbers and boolean values as delimiters in ArrayToList(). The following statements are valid:

```
<cfset lstParms = ArrayToList(aryParms,true)>
<cfset lstParms = ArrayToList(aryParms,0)>
<cfset lstParms = ArrayToList(aryParms,3.14)>
```

Though the preceding examples seem a bit odd, this is the sort of familiarity with CF that I'm trying to encourage – if you know all the ways you could handle a problem, odds are you'll tend to use an appropriate one.

ArrayClear() Function

The ArrayClear() function clears the contents of an array. That is, it returns the array to the same state it was in following an initial ArrayNew() statement. This can be a bit hazardous if you are dealing with a multi-dimensional array. If you perform an ArrayClear() on the overall array, you're effectively transforming it into an empty one-dimensional array. Consider the following code and its output shown below:

```
<cfset aryParms = ArrayNew(3)>
<cfscript>
  aryParms[1][1][1] = "111";
  aryParms[1][1][2] = "112";
  aryParms[1][2][1] = "121";
</cfscript>
<cfdump var="#aryParms#"><br>
<cfset ArrayClear(aryParms)>
<cfdump var="#aryParms#">
```

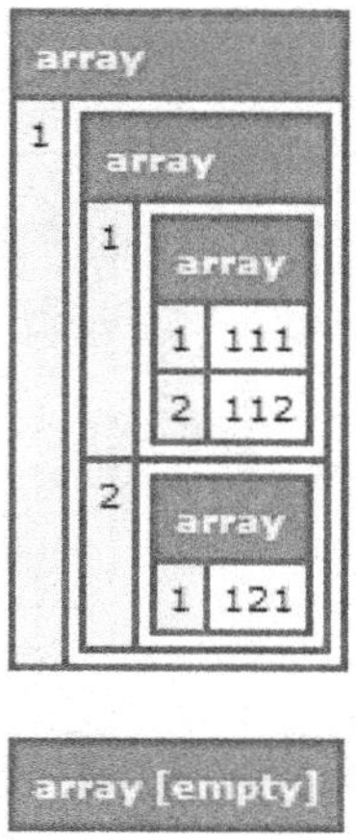

Illustration 7: ArrayClear()

Our nice three-dimensional array was instantly converted into one dimension with no elements. Just something to keep in mind.

ArrayDeleteAt() Function

The ArrayDeleteAt() function removes an element from an array. It takes two arguments: the array to process and an integer that specifies the position of the element to delete. When an element is deleted from within an array, the array closes the gap. Thus, elements to the right (higher indices) of the deleted element will have their index numbers decremented by 1. The following code and illustration show this behavior.

```
<cfset aryTones = ArrayNew(1)>
<cfscript>
 ArrayAppend(aryTones,"do");
 ArrayAppend(aryTones,"re");
 ArrayAppend(aryTones,"mi");
 ArrayAppend(aryTones,"fa");
 ArrayAppend(aryTones,"sol");
 ArrayAppend(aryTones,"la");
 ArrayAppend(aryTones,"ti");
</cfscript>
<cfdump var="#aryTones#">
<cfset ArrayDeleteAt(aryTones,2)>
<cfdump var="#aryTones#">
```

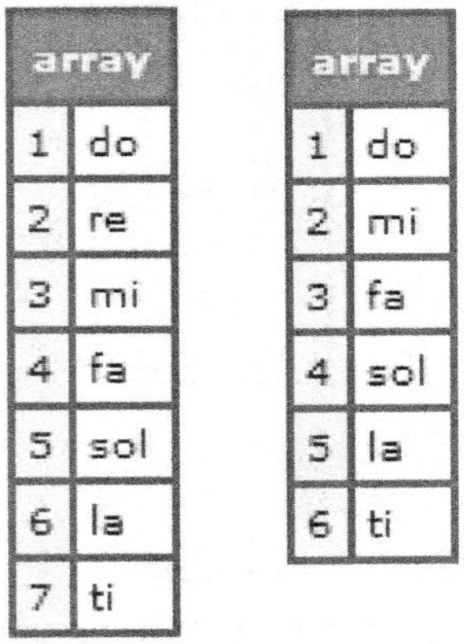

Illustration 8: ArrayDeleteAt()

ArrayMax() Function

The ArrayMax() function returns the greatest value within an array of numeric values. It takes one argument – the array to be processed. If a variable of datatype other than an array is specified, an error is thrown: “Object of type class java.lang.String cannot be used as an array”. If the specified array is of more than one dimension, an error is thrown: “Non-numeric value found” If any element of the array is non-numeric, an error is thrown: “Non-numeric value found”.

ArrayMin() Function

The ArrayMin() function is the converse of the ArrayMax() function – it returns the least value in an array of numeric values. Error messages are the same as for the ArrayMax() function.

ArraySet() Function

The ArraySet() function adds elements to a one-dimensional array. It takes four arguments: the array to process, the starting position, the ending position, and the value to be stored in each of the added elements. It returns true if all the arguments are valid. If the array in the first argument is of more than one dimension, an error is thrown. If the second or third arguments are non-integer values, an error is thrown. If the third value is greater than the size of the array, elements are added to the array.

You can use any datatype for the value to add using ArraySet(). This means that you can use ArrayNew() as the value to create multi-dimensional arrays if you choose. The following code uses loops, ArrayNew() and ArraySet() to create a four-dimensional array and populate each element with an X.

```
<cfset aryStuff = ArrayNew(1)>
<cfscript>
 ArraySet(aryStuff,1,3,ArrayNew(1));
 for (i = 1; i LT 4; i = i + 1) {
  ArraySet(aryStuff[i],1,3,ArrayNew(1));
  for (j = 1; j LT 4; j = j + 1) {
   ArraySet(aryStuff[i][j],1,3,ArrayNew(1));
   for (k = 1; k LT 4; k = k + 1) {
    ArraySet(aryStuff[i][j][k],1,3,"X");
   }
  }
 }
</cfscript>
<cfdump var="#aryStuff#">
```

This code serves to illustrate the flexibility of choices that are available when creating and manipulating arrays in multiple dimensions. You could continue this nesting behavior for as many dimensions as necessary for you purpose.

IsArray() Function

The IsArray() function takes two arguments – an expression that evaluates to a single result and an optional number. If only the first argument is used, the function returns YES if the expression evaluates to an array, and NO if the expression evaluates to any other datatype. If both arguments are used, the function returns YES if the expression evaluates to an array and has the number of elements specified in the second argument, and NO if the expression evaluates to any other datatype or has a different array length than specified in the second argument.

ArraySort() Function

The ArraySort() function arranges the elements of an array in order. It returns yes if the sort is successfully completed, and no if the sort fails (more on this later). It takes three attributes: the array to process, the type of sort to perform, and the direction of sorting ("asc" for ascending or "desc" for descending). If left out, the last argument defaults to ascending.

There are three possible sort types: "numeric", "text" and "textnocase". The first type sorts in numerical order, the second in alphabetical order paying attention to capital letters, and the third in alphabetical order ignoring capital letters. If you attempt to sort an array numerically and the array contains any non-numeric values, an error is thrown.

Adobe's documentation states that ArraySort() returns NO if the sort fails. In practice, though, I have yet to find a way to construct an array in such a way that a sort will fail without throwing an error. So for practical purposes, ArraySort() will always return YES.

ListToArray() Function

The ListToArray() function is the coverse of the ArrayToList() function – it transforms a list into a one-dimensional array. It takes two arguments: the list to process and the list's delimiter(s). If the second argument is left out, it defaults to a comma. If the first argument is of a datatype other than list, an error is thrown. (Keep in mind that a simple value is technically a list with one element).

If a delimiter with more than one character is specified, the function will split the list at each occurrence of each character in the delimiter – NOT each occurrence of the whole delimiter string. To illustrate this, let's reverse the example I used for the ArrayToList() function:

```
<cfset lstParms = "firstName > 'A' AND age < 55 AND gender = 'F'">
<cfset aryParms = ListToArray(lstParms," AND ")>
<cfdump var="#aryParms#">
```

This code results in what may be an unexpected result for you. If you expected the function to strip out occurrences of " AND " and put the remaining pieces into array elements, you get a bit of a shock when you see the actual output:

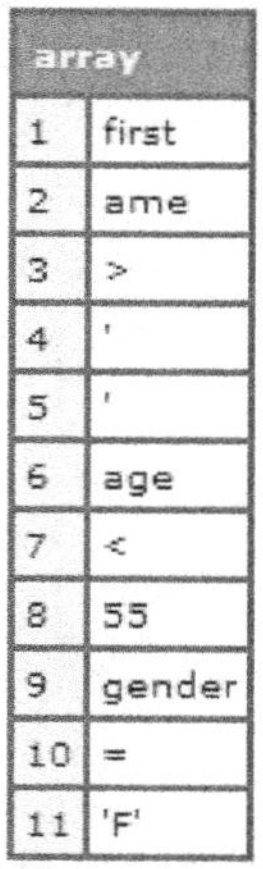

array	
1	first
2	ame
3	>
4	'
5	'
6	age
7	<
8	55
9	gender
10	=
11	'F'

Illustration 9: ListToArray()

The ListToArray() function split up the list on each character of the delimiter string, rather than the delimiter string as a whole. In other words, the list was broken on each space, letter A, letter N, and letter D, resulting in the array shown in Illustration 7.

ArrayInsertAt() Function

The ArrayInsertAt() function inserts an element into an array at the specified position. It takes three arguments: the array to process, a number specifying the position at which to insert the new element, and the value to store in the new element. If the first argument is of a datatype other than array, an error is thrown. If the second argument is of a datatype other than numeric, or is not an integer, or is outside the range of existing index numbers in the array, an error is thrown. If the third argument is omitted, an error is thrown. The function always returns YES. The following code and illustration demonstrate ArrayInsertAt() in action.

```
<cfset aryStuff = ArrayNew(1)>
<cfscript>
 ArrayAppend(aryStuff,"1");
 ArrayAppend(aryStuff,"2");
 ArrayAppend(aryStuff,"3");
 ArrayAppend(aryStuff,"4");
 ArrayInsertAt(aryStuff,3,"Three");
</cfscript>
<cfdump var="#aryStuff#">
```

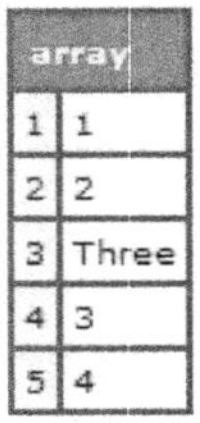

Illustration 10: ArrayInsertAt()

You cannot use ArrayInsertAt() to populate an empty array. Because there are no valid index numbers in an empty array, the function will throw an error if you try to use it on an empty array.

ArrayNew() Function

The ArrayNew() function creates a new empty array. It takes one argument – a number specifying the number of dimensions of the new array (1 to 3). If the argument is omitted, or a value other than 1, 2 or 3 is specified, an error is thrown. The function returns an empty array.

ArrayNew() can cause some interesting results, so it's important to understand how it works. Regardless of the value of the argument specified, the result of the function is an empty array. If you <CFDUMP> the result, you can't tell the difference between the result of ArrayNew(1), ArrayNew(2), and ArrayNew(3). But there is an invisible difference. The argument allows ColdFusion to define the array in the proper number of dimensions. So, the following code results in a two-dimensional array with some values in it.

```
<cfset aryMine = ArrayNew(2)>
<cfset aryMine[1][1] = 93>
<cfset aryMine[1][2] = 100>
```

But what happens if we get the argument wrong? Consider the following code:

```
<cfset aryMine = ArrayNew(1)>
<cfset aryMine[1][1] = 93>
<cfset aryMine[1][2] = 100>
```

You might be surprised to know that this code runs just fine. ColdFusion allows the syntax in the second and third lines, even though the array was defined in only one dimension. However, CF doesn't store the values in a two-dimensional array. Instead, it creates a structure inside the one-dimensional array and stores the values in the structure. (I'll get into why this happens in the next chapter – Structures.) The illustration shows the difference in the results from the two blocks of code.

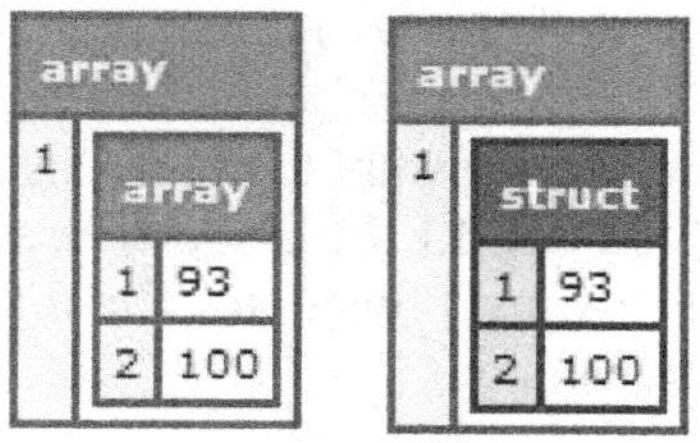

Illustration 11: ArrayNew() with Wrong Dimensions Argument

ArraySum() Function

The ArraySum() function returns the sum of all elements of an array of numeric values. It takes one argument – the array to be processed. If the array contains anything other than numeric values, an error is thrown. If the variable passed as the argument is of a datatype other than array, an error is thrown.

Now that we've had a look at managing both one-dimensional and multi-dimensional arrays, and ColdFusion's array functions, we'll run through a few quick examples that might come in handy in your web development work.

Array Examples

The previous how-to sections have provided a number of generic array examples. The following sections are aimed specifically at some practices for web development. You'll probably even find uses for some of them on your latest project.

Array Example 1: Create An Array of Form Field Names

In web development, one of the most often-performed tasks is processing an HTML form. There are probably as many methods of processing form data in existence as there are development shops. Regardless of the method you use, it's helpful to be able to process the fields on a form as a group. You can do this by treating the set of form fields as a list, an array, or even a structure. In fact, the FORM scope in ColdFusion MX is actually a structure – something we'll look at in the next chapter.

But let's start with the assumption that we want to put the list of form field names in an array. The list is available in the variable FORM.FIELDNAMES, so it's a simple matter to convert the list to an array using the ListToArray() function. The following code shows the form and an action page that handles the conversion.

```
<!--- dspFieldsToArray.cfm --->
<form action="actFieldsToArray.cfm" method="post">
First name: <input name="firstName"><br>
Last name: <input name="lastName"><br>
<input type="submit">
</form>
<!--- actFieldsToArray.cfm --->
<cfset aryFieldNames = ListToArray(form.fieldNames)>
<cfdump var="#aryFieldNames#">
```

The illustration below shows the form and the result of submitting it. Note that we're not dealing with form field values yet – that's the job of the next example. Also, the submit button was not passed as a field; that's because we didn't assign it a name attribute.

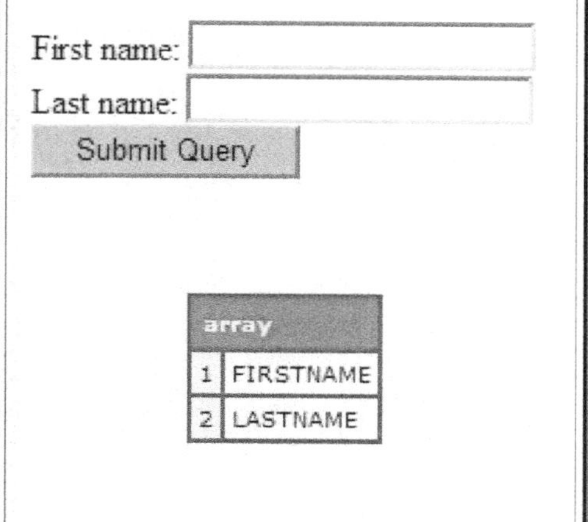

Illustration 12: Form Array

Array Example 2: Create An Array of Form Field Values

Where the previous example developed an array of field names from a form, this example assumes we need to get the values from the fields into an array. We'll also add a name and value to the submit button to see how that affects things.

To get the form field values into an array, we need to loop over the list of field names and get the value from the form structure to store into the array.

```
<!--- dspFieldValuesToArray.cfm --->
<form action="actFieldValuesToArray.cfm" method="post">
First name: <input name="firstName"><br>
Last name: <input name="lastName"><br>
<input type="submit" name="btnSave" value="Save">
</form>
<!--- actFieldValuesToArray.cfm --->
<cfset aryFieldValues = ArrayNew(1)>
<cfloop from="1" to="#ListLen(form.fieldNames)#" index="i">
  <cfset aryFieldValues[i] = form[ListGetAt(form.fieldNames,i)]>
</cfloop>
<cfdump var="#aryFieldValues#">
```

The next illustration shows the form and the result of submitting it.

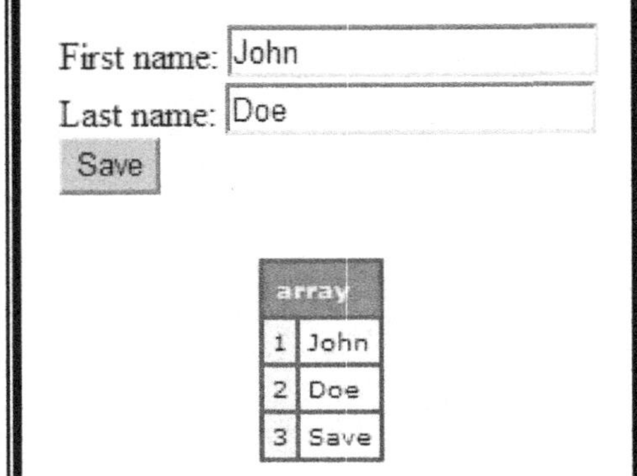

Illustration 13: Completed Form Array

Array Example 3: Sort Names

Arrays are a great means to sort data. For simple sorting, the ArraySort() function is the way to go. When sorting names, we want to use an alphabetical sort, indicated in the ArraySort() function with the "text" argument. The following code sets up an array of names, displays the array, sorts and array, and displays it again. The output is shown in the illustration.

```
<cfset aryNames = ArrayNew(1)>
<cfset ArrayAppend(aryNames, "Delta, Derrick")>
<cfset ArrayAppend(aryNames, "Echo, Ernestine")>
<cfset ArrayAppend(aryNames, "Baker, Barry")>
<cfset ArrayAppend(aryNames, "Able, Anita")>
<cfset ArrayAppend(aryNames, "Charlie, Carl")>
<cfdump var="#aryNames#">
<cfset ArraySort(aryNames, "text")>
<cfdump var="#aryNames#">
```

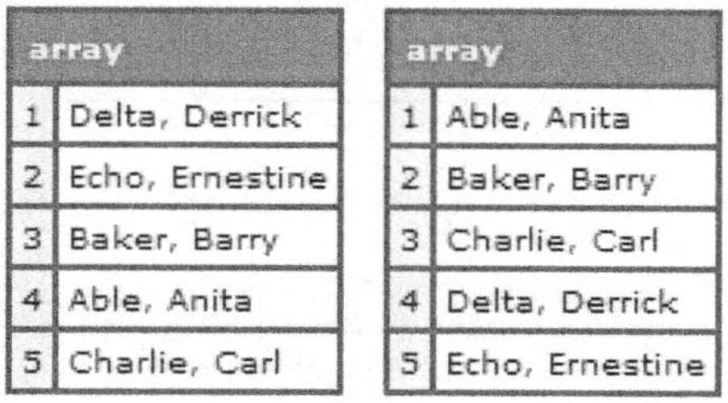

array	
1	Delta, Derrick
2	Echo, Ernestine
3	Baker, Barry
4	Able, Anita
5	Charlie, Carl

array	
1	Able, Anita
2	Baker, Barry
3	Charlie, Carl
4	Delta, Derrick
5	Echo, Ernestine

Illustration 14: Sorted Array

Array Example 4: Advanced Sorting

ArraySwap() is a good way to create sort routines other than the basics provided by the ArraySort() function. The previous example showed how easy it is to create simple sorts. This example is a bit more complicated. It reverses the format of each name. Now, instead of being in lastName, firstName format, they are in firstName lastName order. We still want to sort by last name, though (and I've mixed up the first and last names to show that the sort doesn't take place on first name). This seems impossible considering the abilities of ArraySort(). With a bit of thinking, though, it's quite simple with ArraySwap(). Here's the code, and its output.

```
<cfset aryNames = ArrayNew(1)>
<cfset ArrayAppend(aryNames, "Barry Delta")>
<cfset ArrayAppend(aryNames, "Anita Echo")>
<cfset ArrayAppend(aryNames, "Derrick Baker")>
<cfset ArrayAppend(aryNames, "Ernestine Able")>
<cfset ArrayAppend(aryNames, "Carl Charlie")>
<cfdump var="#aryNames#"><br>

<cfloop from="1" to="#ArrayLen(aryNames)#" index="iteration">
  <cfloop from="#ArrayLen(aryNames)#" to="2" step="-1" index="i">
    <cfif ListLast(aryNames[i], " ") LT ListLast(aryNames[i-1], " ")>
      <cfset ArraySwap(aryNames,i,i-1)>
    </cfif>
  </cfloop>
</cfloop>
<cfdump var="#aryNames#">
```

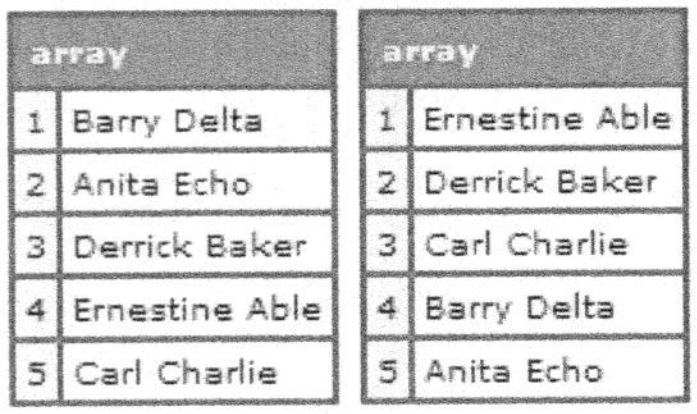

Illustration 15: Sorted Array

This is an example of a "brute force" bubble sort. There are many other sort algorithms you can implement, but they're beyond the scope of this book.

Array Example 5: Stacks (Push and Pop)

Stacks are a frequently employed means of handling data. A stack is a construct where the last piece of data placed on the stack is the first piece of data retrieved. You can think of a stack as one of those spring-loaded plate dispensers you've probably seen at a buffet. New plates are put on the stack from the top (the server pushes them down onto the stack), and when you need one you take the one off the top. You can't take any plate except the one on the top.

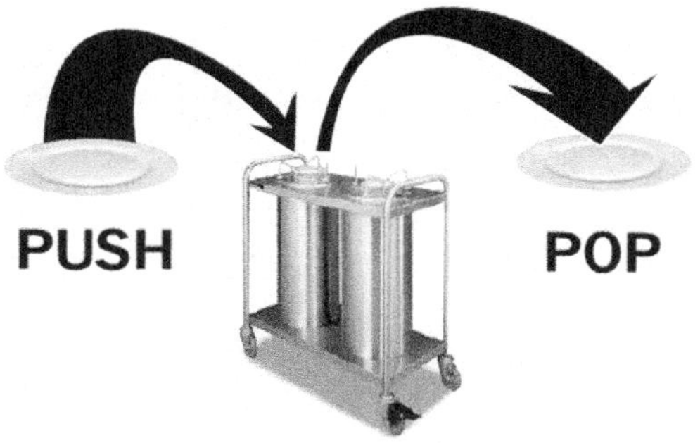

Illustration 16: A Physical Stack

This is exactly how stacks work in programming. When you put a value on the stack, it's called pushing a value. When you get a value from the stack, it's called popping a value.

Arrays make very good stacks. You can use ArrayPrepend() to push values, and ArrayDeleteAt() to remove the first value when you pop it. These tasks are well suited for User Defined Functions (UDFs), as shown below.

```
<!--- stackLib.cfm --->
<cfscript>
function stackPush(aryStack, varItem) {
 ArrayPrepend(aryStack, varItem);
 return aryStack;
}

function stackPop(aryStack) {
 <!--- poppedItem must exist in the calling page before stackPop() is called --->
 poppedItem = aryStack[1];
 ArrayDeleteAt(aryStack, 1);
 return aryStack;
}
</cfscript>

<!--- stackExample.cfm --->
<cfinclude template="stackLib.cfm">
<cfset aryPlates = ArrayNew(1)>
Here's how ArrayAppend creates elements<br>
<cfscript>
 ArrayAppend(aryPlates, "Blue");
 ArrayAppend(aryPlates, "Upper");
 ArrayAppend(aryPlates, "Chrome");
</cfscript>
<cfdump var="#aryPlates#">
<br>
And here's how stackPush creates elements<br>
<cfscript>
 ArrayClear(aryPlates);
 aryPlates = stackPush(aryPlates, "Blue");
 aryPlates = stackPush(aryPlates, "Upper");
 aryPlates = stackPush(aryPlates, "Chrome");
</cfscript>
<cfdump var="#aryPlates#">
<br>
Finally, we'll pop a plate and push a new one<br>
<cfscript>
 poppedItem = "";
 aryPlates = stackPop(aryPlates);
 aryPlates = stackPush(aryPlates, "Dresden");
</cfscript>
<cfoutput>
I popped a #poppedItem# and now the stack looks like this<br>
</cfoutput>
<cfdump var="#aryPlates#">
```

The illustration below shows the stack demonstration in action. Stacks are useful anytime you need to keep track of a last in, first out (LIFO) collection of values. You can also store complex datatypes in a stack. Lastly, stacks are related to queues, which are simply first in, first out (FIFO) collections. The actions related to a queue are usually called queue (analogous to push) and dequeue (analogous to pop). I'll leave you to create your own queue management UDFs, using the stack UDFs above as a starting point.

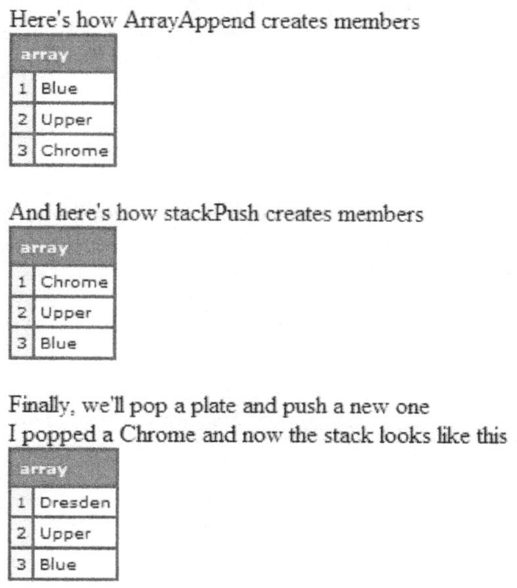

Illustration 17: Stack in Action

Array Example 6: A Deck of Cards

For the final array example, we'll take a look at how we might manage a deck of cards using an array. There are many ways we might want to manipulate the deck, and an array will let us handle them. Look over the example code carefully. I've used two-letter abbreviations for the suits (CLubs, HEarts, DIamonds, and SPades) to save space in the output, and numbers for the face cards (11=Jack, 12=Queen, 13=King) to make the code simple.

```
<cfset aryDeck = ArrayNew(1)>
<!--- Load the deck --->
<cfset lstSuits = "CL,HE,DI,SP">
<cfloop list="#lstSuits#" index="thisSuit">
  <cfloop from="1" to="13" index="cardValue">
    <cfset ArrayAppend(aryDeck, "#cardValue##thisSuit#")>
  </cfloop>
</cfloop>
Here's the new deck:
<cfdump var="#aryDeck#">
<br>

<!--- Shuffle the deck --->
<!--- This works like a bubble sort, but uses a randomizer to decide the target of each swap --->
<cfloop from="1" to="#ArrayLen(aryDeck)#" index="i">
  <cfscript>
    cardToSwapWith = 0;
    while (cardToSwapWith EQ 0) {
      cardToSwapWith = Int(Rand()*52);
    }
  </cfscript>
  <cfset ArraySwap(aryDeck,i,cardToSwapWith)>
</cfloop>
And here's the shuffled deck:
<cfdump var="#aryDeck#"><br>

<!--- Deal the cards --->
<cfset numberOfHands = 4>
<cfset cardsPerHand = 5>
<cfset aryHands = ArrayNew(1)>
<cfset temp = ArraySet(aryHands,1,4,"")>
<cfset thisHand = 1>
<cfloop from="1" to="#numberOfHands * cardsPerHand#" index="i">
  <cfscript>
    aryHands[thisHand] = ListAppend(aryHands[thisHand],aryDeck[1]);
    ArrayDeleteAt(aryDeck,1);
    thisHand = thisHand + 1;
    if (thisHand EQ 5) {
      thisHand = 1;
    }
  </cfscript>
</cfloop>
<cfoutput>
Here are #numberOfHands# hands dealt out:
</cfoutput>
<cfdump var="#aryHands#">
And the remaining deck:
<cfdump var="#aryDeck#">
```

And after all that, we'll take a look at the output.

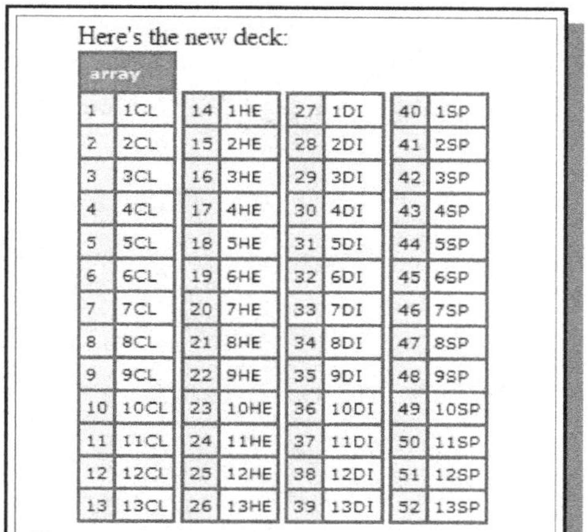

Here's the new deck:

array

1	1CL	14	1HE	27	1DI	40	1SP
2	2CL	15	2HE	28	2DI	41	2SP
3	3CL	16	3HE	29	3DI	42	3SP
4	4CL	17	4HE	30	4DI	43	4SP
5	5CL	18	5HE	31	5DI	44	5SP
6	6CL	19	6HE	32	6DI	45	6SP
7	7CL	20	7HE	33	7DI	46	7SP
8	8CL	21	8HE	34	8DI	47	8SP
9	9CL	22	9HE	35	9DI	48	9SP
10	10CL	23	10HE	36	10DI	49	10SP
11	11CL	24	11HE	37	11DI	50	11SP
12	12CL	25	12HE	38	12DI	51	12SP
13	13CL	26	13HE	39	13DI	52	13SP

Illustration 18: The Deck

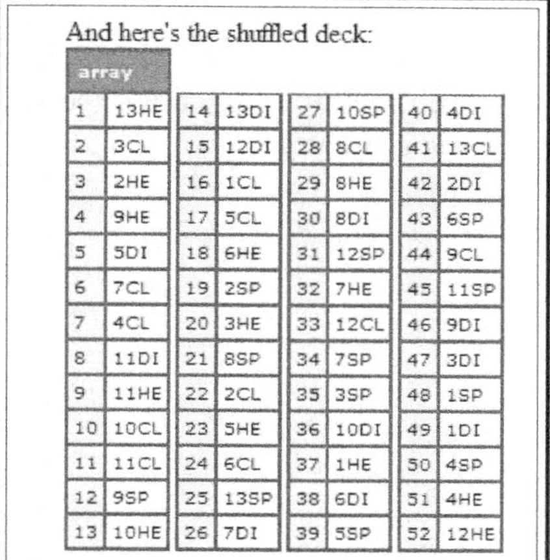

And here's the shuffled deck:

array

1	13HE	14	13DI	27	10SP	40	4DI
2	3CL	15	12DI	28	8CL	41	13CL
3	2HE	16	1CL	29	8HE	42	2DI
4	9HE	17	5CL	30	8DI	43	6SP
5	5DI	18	6HE	31	12SP	44	9CL
6	7CL	19	2SP	32	7HE	45	11SP
7	4CL	20	3HE	33	12CL	46	9DI
8	11DI	21	8SP	34	7SP	47	3DI
9	11HE	22	2CL	35	3SP	48	1SP
10	10CL	23	5HE	36	10DI	49	1DI
11	11CL	24	6CL	37	1HE	50	4SP
12	9SP	25	13SP	38	6DI	51	4HE
13	10HE	26	7DI	39	5SP	52	12HE

Illustration 19: Shuffled Deck

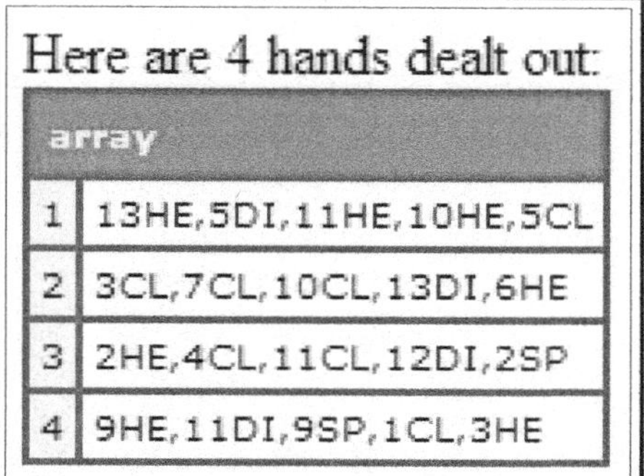

Here are 4 hands dealt out:

array

1	13HE,5DI,11HE,10HE,5CL
2	3CL,7CL,10CL,13DI,6HE
3	2HE,4CL,11CL,12DI,2SP
4	9HE,11DI,9SP,1CL,3HE

Illustration 20: Dealt Hands

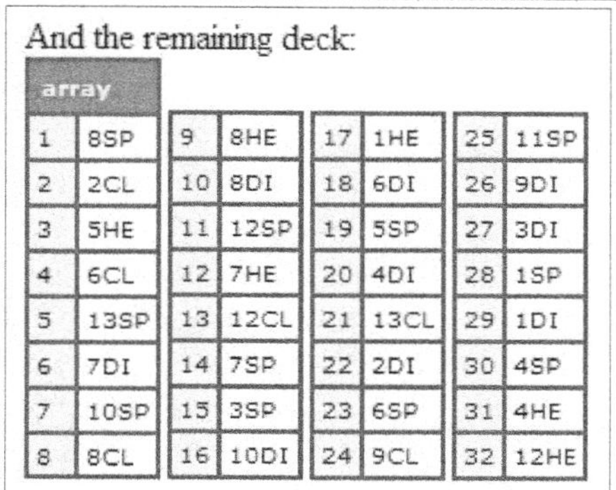

And the remaining deck:

array

1	8SP	9	8HE	17	1HE	25	11SP
2	2CL	10	8DI	18	6DI	26	9DI
3	5HE	11	12SP	19	5SP	27	3DI
4	6CL	12	7HE	20	4DI	28	1SP
5	13SP	13	12CL	21	13CL	29	1DI
6	7DI	14	7SP	22	2DI	30	4SP
7	10SP	15	3SP	23	6SP	31	4HE
8	8CL	16	10DI	24	9CL	32	12HE

Illustration 21: Remaining Deck

Arrayed Out Before Us

Arrays are very useful and flexible constructs. They can be used to solve a wide variety of programming problems where you need to manage a collection of values that have no identifying features. This chapter has stepped through all the functions you can employ to manipulate arrays in ColdFusion, and provided a number of examples to help you get a handle on them.

So we've covered lists and arrays. We're two-thirds of the way through the major topics of this book. Now we're ready to move on to a datatype that is related to arrays but treated much differently – the structure.

3.Structures

What Is a Structure?

In ColdFusion, the word structure refers to a specialized complex datatype. Structures are related to arrays in that they are a way to store a collection of information in a single variable. However, structures have a different means of addressing the members of the collection. Where an array's elements are referenced by index numbers, a structure's elements are referenced by a key string associated with each element. For this reason, the construct that ColdFusion calls a structure is often referred to in other languages as an "associative array". Each value in the construct is referred to by its unique key.

Since a structure is comprised of these name/value pairs, there can be no such thing as a multi-dimensional structure. Structures exist in only one form. This makes the decision process for using a structure fairly simple. Any time you need to store a collection of data where each piece of data can be referenced by a name, a structure is a good way to go. Note, though, that structures can be nested. We'll see more on that idea later.

When data is stored in a structure, two pieces of information are needed: the datum to be stored and the key by which it will be referred. For example, you might create a structure called Student to hold data about a student. The structure might hold values such as the student's ID number, first and last names, and dorm assignment. The name/value pairs for the Student structure could be represented like this:

```
[Student]

student ID=123456789

last Name=Smith

first Name=Sarah

dorm=Gooch
```

This format might look familiar to you; it is the Microsoft Windows .INI file format. The "structure" is represented by a section header, each key is on the left side of an equals sign, and each value is on the right side of an equals sign, opposite its corresponding key.

Based on this layout, we can easily retrieve any value by supplying the section and key names. In ColdFusion, the section is equivalent to the structure, and the key is still the key. We could ask for the firstName key of the Student structure by asking for Student.firstName. We'll look at structure reference syntax more closely in the following sections. First we'll take a step-by-step look at using structures.

Structures How-To

Because structures are not subject to the kind of multi-dimensional complexity available to arrays, we don't need to spend nearly as much time looking at how they work. In the following sections we'll explore how to create a structure, add elements to it, remove elements from it, sort the elements in it, and loop over the elements to operate on each one.

Create a Structure

Just as with arrays and the ArrayNew() function, structures are created in ColdFusion with the StructNew() function. Unlike ArrayNew(), though, StructNew() takes no arguments. It simply creates and returns a new empty structure. The following statements are equivalent:

```
<cfset Student = StructNew()>
<cfscript>Student = StructNew();</cfscript>
```

Starting with ColdFusion MX, a new and subtle structure creation process is available. If you refer to a variable using a structure-access syntax, CFMX will create the structure if it doesn't already exist. Consequently, if the Student structure doesn't exist, the following statement creates a new structure and adds a key/value pair to it:

```
<cfset Student.studentID = "123456789">
```

This is why the example in the last chapter about referencing non-existent array dimensions works the way it does. For clarity's sake, we'll repeat the example here.

```
<cfset aryStuff = ArrayNew(1)>
<cfset aryStuff[1][1] = "foo">
<cfset aryStuff[1][2] = "bar">
```

The following example includes an error in the number of dimensions in the ArrayNew() declaration as compared to the CFSET statement in the following line. However, the CFSET doesn't throw an error. Instead, it creates a structure inside the one-dimensional array. Why does this happen? Because there are two valid ways to reference the elements of a structure.

We've already covered the first way to reference a value in a structure: dotted syntax. In the example above, we used Student.studentID to refer to the value of studentID in the structure named Student.

We could refer to the same element of the Student structure with the following:

```
Student["studentID"]
```

This alternate syntax allows us to use a variable to reference the element, enabling the use of loops and other variable means to reuse code. (More on loops and structures later.) It also means that CFMX's handy auto-structure-creation service reacts equally well to this syntax as it does to the dotted syntax. Add in the fact that, since CF variables can't begin with numbers, CFMX will interpret a number inside the square brackets as a literal, and the recipe is complete. The following syntax causes the creation of a structure named Stuff (if it doesn't already exist), and populates the structure with a key named 1 whose value is "baseball":

```
<cfset Stuff[1] = "baseball">
```

All these factors working together mean that the statement in the example above:

```
<cfset aryStuff[1][1] = "foo">
```

results in the creation of a new structure named aryStuff[1], with a key named "1". This is how we ended up with a structure inside an array. Of course, if the first line of the example had used a correct argument for the ArrayNew() function, we would have had a two-dimensional array, as illustrated in the last chapter. Beware those arguments!

Now that we're familiar with the ins and outs of creating a structure, we need to be able to work with data within it. So let's add some data to the structure.

Add Members To a Structure

As noted previously, adding elements to a structure is a very simple operation. All we need to do is name the element with a key, and assign it a value. The two syntaxes we can use are dotted syntax and bracketed syntax, shown respectively in the following lines of code.

```
<cfset stuStudent.studentID = “123456789”>
<cfset stuStudent[“studentID”] = “123456789”>
```

The bracketed syntax is useful for operating on several keys in a structure with one piece of code. For example, if we need to initialize the Student structure with empty strings, we can do it very easily with the following code:

```
<cfset lstStudentFields = "studentID,firstName,lastName,dormitory">
<cfset stuStudent = StructNew()>
<cfloop list="#lstStudentFields#" index="thisItem">
  <cfset stuStudent[thisItem] = "">
</cfloop>
<cfdump var="#stuStudent#">
```

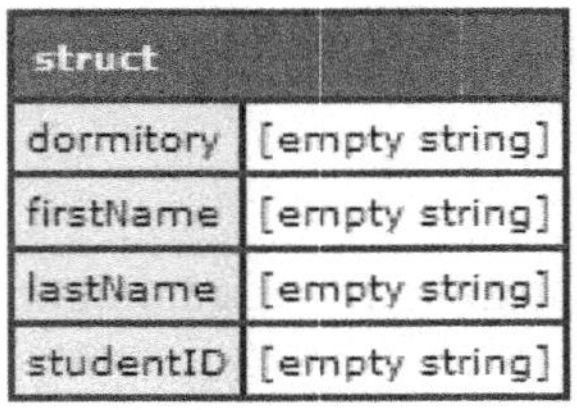

struct	
dormitory	[empty string]
firstName	[empty string]
lastName	[empty string]
studentID	[empty string]

Illustration 22 - New Structure

The result of running this code is shown in the illustration above. It would be just as easy to create the same structure by hard-coding a <CFSET> for each field, but then the code couldn't be easily adapted to adding a new field, or using field names from a form or database. Flexibility is better.

Now that we've added the elements to our stuStudent structure, let's toss in some useful data to use for the examples in the coming sections. We'll make the structure represent student ID **123456789**, which is for **Sarah Smith**, a student assigned to **Gooch** dormitory. The code and output are shown below:

struct	
DORMITORY	Gooch
FIRSTNAME	Sarah
LASTNAME	Smith
STUDENTID	123456789

Illustration 23 - Populated Structure

```
<cfscript>
  stuStudent.studentID = "123456789";
  stuStudent.lastName = "Smith";
  stuStudent.firstName = "Sarah";
  stuStudent.dormitory = "Gooch";
</cfscript>
```

With the introduction of ColdFusion 8, we have one more variation available for creating a structure. This approach creates the structure and populates several keys all in one statement. Here's an example using our Student structure:

```
<cfset stuStudent = {studentID=123456789,
lastName="Smith",firstName="Sarah",dormitory="Gooch"}>
```

This newest syntax makes structure creation much easier from a coding point of view. The result of this statement is the structure seen in the illustration above.

If we can add elements to a structure, odds are that sooner or later we'll need to remove one or more of them. CF makes this just as easy.

Remove Members From a Structure

To remove an element from a structure, first we need to make sure we know what we mean by the term “remove”. Every value in a structure is stored under its key. This means we can remove either of two distinct things: the value, or the entire name/value pair.

Removing the value of an element is simple; we can just set the value of the element to an empty string. For example, we can set the value of dormitory in the example structure to an empty string to indicate that Sarah doesn't have a room assignment:

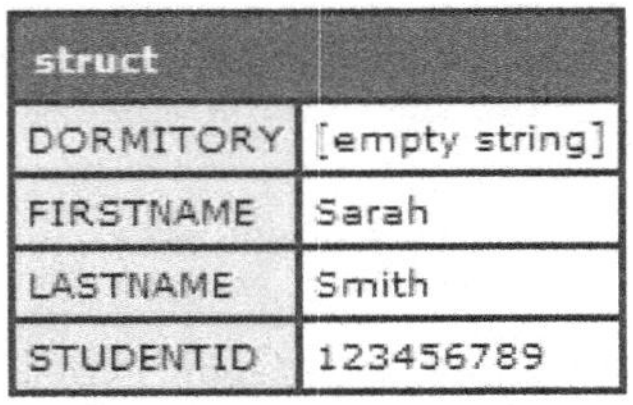

struct	
DORMITORY	[empty string]
FIRSTNAME	Sarah
LASTNAME	Smith
STUDENTID	123456789

Illustration 24 - Empty Dormitory

```
<cfset stuStudent.dormitory = “”>
```

On the other hand, maybe we want to remove the entire dormitory name/value pair. This would be accomplished by using the StructDelete() function, as shown below.

```
<cfset StructDelete(stuStudent, "dormitory")>
```

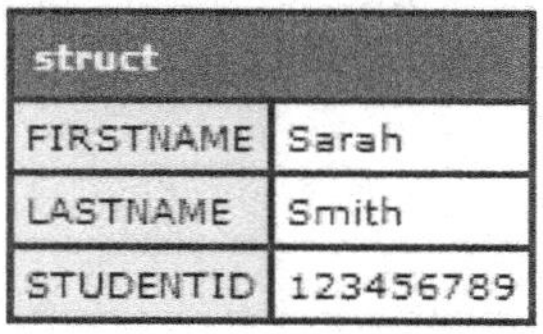

Illustration 25 - Dormitory Deleted

There is a third optional argument for the StructDelete() function; it is discussed in detail in the section on that function later in this chapter.

Sort a Structure

It probably comes as no surprise that structures can be sorted using the StructSort() function. But before we jump off into sorting structures, let's think about it for a minute. What's the point of sorting a structure anyway? After all, we read data from a structure based on key values. It doesn't matter where in the structure any given key is; we can still read it whenever we want for whatever purpose we want. So why even have a StructSort() function?

The answer is so we can return sorted information about a structure, as opposed to sorting the structure itself. In fact, the StructSort() doesn't sort a structure. Rather, it returns an array of the keys in a structure sorted in terms of their values. For example, if we had a structure that contained product numbers as its keys, and prices as its values, we could get an array with the product numbers sorted from least expensive to most expensive with the following command:

```
<cfset aryProductsByPrice = StructSort(stuProducts)>
```

The result doesn't contain any price information – just the product numbers in order of price. If we use nested structures, StructSort() can be used to perform sub-sorts and sub-sub-sorts as well.

Although sorts are not terribly common practice where structures are concerned, looping over a structure is a very common practice. The next section discusses structure-based loops.

Loop Over a Structure

Structures are exceptionally handy constructs where looping is concerned. With a well-constructed loop, we can easily process a piece of code on every element of the structure. Add to this the fact that in CFMX most of the scopes are now stored as structures, and you have a exceptionally easy way to manage variable data. For example, if we need a template to tell us all the variables that are in the form scope and their current values, we can create it with a simple loop:

```
<cfif not isDefined("form.btnSubmit")>
  <form name="myForm" method="post">
    <input type="hidden" name="department" value="Philosophy">
    User name: <input type="text" name="userName"><br>
    <input name="btnSubmit" type="submit">
  </form>
<cfelse>
  <cfloop collection="#form#" item="thisItem">
    <cfoutput>form.#thisItem# = #form[thisItem]#<br></cfoutput>
  </cfloop>
</cfif>
```

Note that this page calls itself (a form with no action specified will call the same template), and processes the output based on the presence of a variable created by the form (form.btnSubmit). The form and the results of the process are shown below.

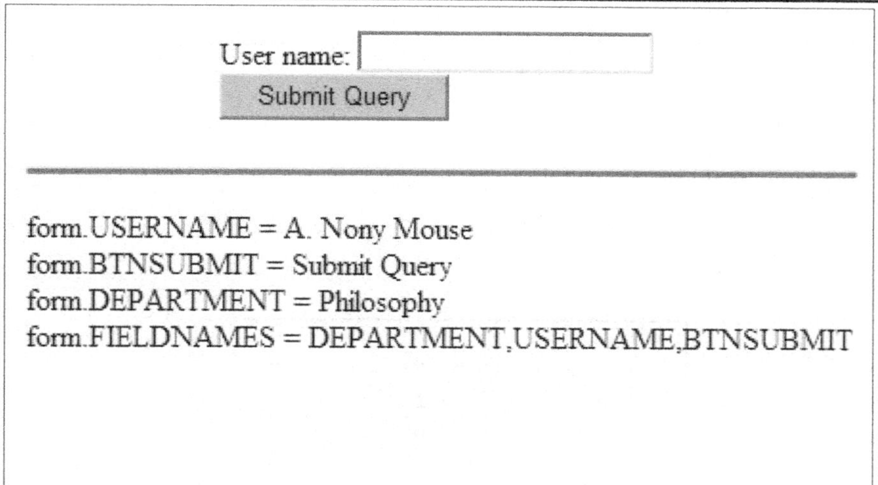

Illustration 26 - Process Form Variables

So with one brief loop we can easily process every element of a structure. This prevents the need to change the code if the content of the structure changes. We'll look at more of ColdFusion's scope structures a little bit later.

As we've seen, structures are quite simple to manage, being so straightforward in design. There's a small complication to consider, though - that of nested structures.

Nested Structures

A nested structure is somewhat like a nested array, in that it's simply a structure full of structures. This might seem like a confusing idea at first, but it's quite straightforward. When nesting elements of a nested structure, we simply extend the dotted notation. The following code sample illustrates nested structures.

```
<cfset Students = StructNew()>
<cfset Students.Student001 = StructNew()>
<cfset Students.Student001.lastName = "Public">
<cfset Students.Student001.firstName = "John">
<cfset Students.Student001.midInitial = "Q">
<cfset Students.Student002 = StructNew()>
<cfset Students.Student002.lastName = "Doe">
<cfset Students.Student002.firstName = "Jane">
<cfset Students.Student002.midInitial = "NMI">

<cfdump var="#Students#">
```

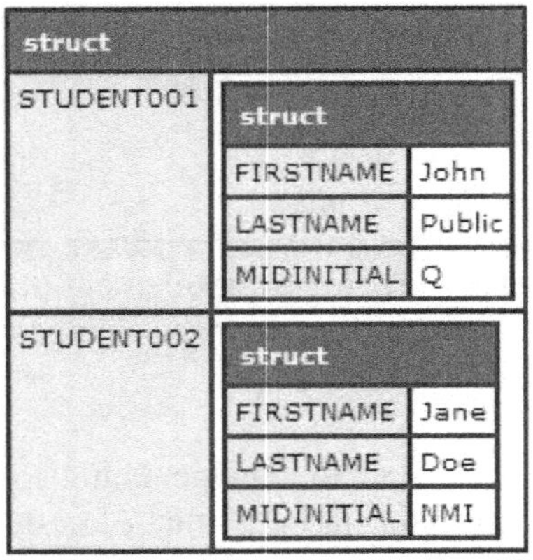

Illustration 27 - Nested Structures

The extensible nature of the dotted notation is apparent in the lines that set lastName, firstName, and midInitial for each student. The same idea is used to reference these elements when retrieving values. To get the middle initial for Student002, we would use #Students.Student002.midInitial#.

We can also use the bracketed notation to reference nested structure elements, although this approach is a little uglier than the dotted notation. The following references are valid for elements of the structure shown in the previous illustration:

```
Students["Student001"]["lastName"]
Students["Student002"].lastName
Students.Student001["lastName"]
```

Why is it useful to have both dotted and bracketed notations available? Consider the following code:

```
<cfloop from="1" to="2" index="i">
  <cfoutput>#Students.Student00#i#.lastName#<br></cfoutput>
</cfloop>
```

The intent here is to loop over some set of students and return each student's last name. But the syntax shown is not valid – we can't nest pound signs that way to allow use of the index variable `i`. We could solve the problem by using the `Evaluate()` function, but that's generally a last-resort option, since `Evaluate()` is a comparatively processor-intensive function.

The solution is to use bracketed notation, which expects a string inside the brackets to reference a key:

```
<cfloop from="1" to="2" index="i">
  <cfset studentID = "Student00" & i>
  <cfoutput>#Students[studentID].lastName#<br></cfoutput>
</cfloop>
```

or, if you prefer:

```
<cfloop from="1" to="2" index="i">
  <cfset studentID = "Student00" & i>
  <cfoutput>#Students[studentID]["lastName"]#<br></cfoutput>
</cfloop>
```

Either of these methods works fine.

PITFALL: Notice in the second example that `lastName` is in quotes in the third line. If it weren't, ColdFusion would look for a variable named `lastName` and, not finding it, throw an error. This is one of the most common errors with bracketed notation, so watch out for it.

There is one other major difference between dot notation and bracket notation. When using dot notation, keys may not be numeric, start with numbers, or contain spaces. These conditions are allowable when using bracket notation. However, most programmers avoid using these special conditions because of the restrictions they impose.

OK, so now with the basics of structure management behind us, we can look into the wide variety of functions that ColdFusion provides for structures.

ColdFusion Structure Functions

Just as most of the array functions start with "Array", so most of the structure functions start with "Struct". The big exception to this is the first function (alphabetically) – IsStruct().

IsStruct()

The IsStruct() function take a single argument – a variable. If the variable in question is a structure, the function returns YES; otherwise it returns NO. If the argument is omitted, an error is returned.

PITFALL: If you wrap the argument in quotes, IsStruct() will always return NO because the argument is a string. Make sure that the expression inside the function returns a variable, NOT the name of the variable.

StructAppend()

The StructAppend() function adds the keys from one structure onto another structure. It takes three arguments: the structure to get the keys from, the structure to which the keys will be added, and a boolean to indicate whether matching keys should have their values overwritten. StructAppend() always returns YES. If either of the first two arguments are not structures, an error is thrown. If the last argument is not a boolean, an error is thrown.

This function is the easiest way to combine the contents of two structures. The following code demonstrates, and the illustration shows the output, first with overwrites allowed and then with overwrites disallowed.

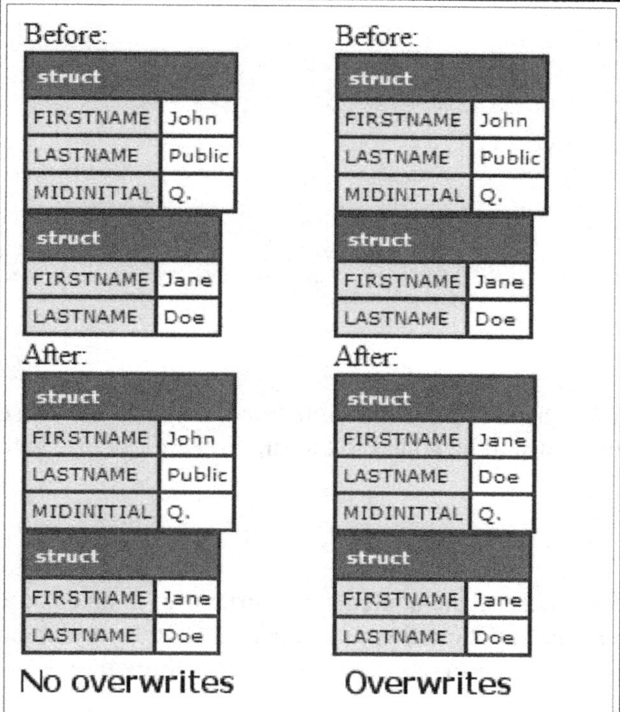

Illustration 28 - StructAppend()

```
<cfset Students = StructNew()>
<cfset Students.firstName = "John">
<cfset Students.lastName = "Public">
<cfset Students.midInitial = "Q.">
<cfset Profs = StructNew()>
<cfset Profs.firstName = "Jane">
<cfset Profs.lastName = "Doe">
Before:
<cfdump var="#Students#">
<cfdump var="#Profs#">
<cfset StructAppend(Students,Profs,"NO")> <!--- Change to YES for second --->
After:
<cfdump var="#Students#">
<cfdump var="#Profs#">
```

The two structures do not need to have the same keys; keys that are present in the second structure but not in the first will be added to the first, regardless of whether overwrites are allowed.

StructClear()

The StructClear() function deletes the contents of a structure. It takes one argument – the structure to be cleared. The function always returns YES. If the argument is omitted, or a variable of type other than structure is passed, an error is thrown.

```
<cfset Students = StructNew()>
<cfset Students.firstName = "John">
<cfset Students.lastName = "Public">
<cfset Students.midInitial = "Q.">
Before:
<cfdump var="#Students#">
<cfset Profs = StructClear(Students)>
After:
<cfdump var="#Students#">
```

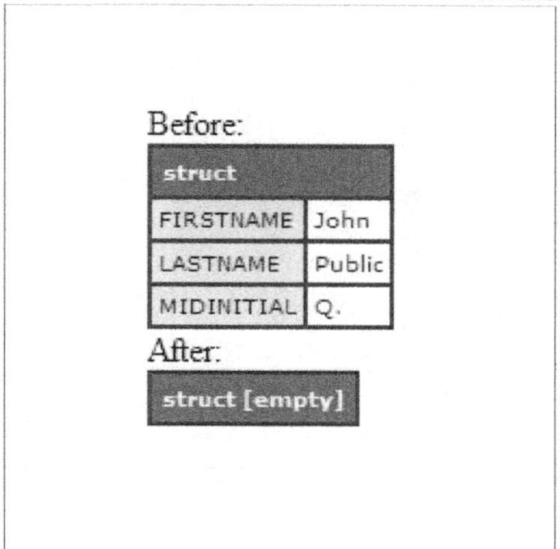

Illustration 29 - StructClear()

StructCopy()

The StructCopy() function copies the contents of one structure into a new structure. It takes one argument: the structure from which to copy. It returns a copy of the specified structure. If the argument is of a type other than structure, an error is thrown.

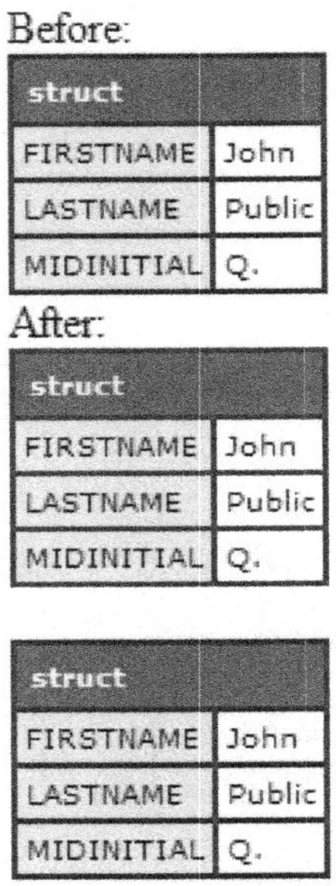

Illustration 30 - StructCopy()

```
<cfset Students = StructNew()>
<cfset Students.firstName = "John">
<cfset Students.lastName = "Public">
<cfset Students.midInitial = "Q.">
Before:
<cfdump var="#Students#">
<cfset People = StructCopy(Students)>
After:
<cfdump var="#Students#"><br>
<cfdump var="#People#">
```

PITFALL: There is a problem with the StructCopy() function. The StructCopy() function does not properly copy a nested structure. Rather, it creates a pointer to the original structure. This is a problem if you create a copy of a structure, then change the value of a key in the original. The value for the same key also changes in the copy, because

the copy is simply a pointer to the original. The Duplicate() function was created to provide a solution to this problem without causing more problems for developers who knew about StructCopy()'s pointer-based behavior and leveraged it in their programs. Duplicate() works just the way StructCopy() does, but it creates a legitimate copy of a nested structure, instead of a pointer.

StructCount()

The StructCount() function returns the number of elements in the specified structure. It takes one argument: the structure to be processed. If the argument is of type other than array, an error is thrown.

```
<cfset Students = StructNew()>
<cfset Students.Student1 = StructNew()>
<cfset Students.Student1.firstName = "John">
<cfset Students.Student1.lastName = "Public">
<cfset Students.Student1.midInitial = "Q.">
<cfset Students.Student2 = StructNew()>
<cfset Students.Student2.firstName = "Jane">
<cfset Students.Student2.lastName = "Doe">
<cfset Students.Student2.midInitial = "NMI">

<cfset numberOfKeys = StructCount(Students)>
<cfdump var="#numberOfKeys#"><br>  <!--- Returns 2 --->
```

As shown in the above code, an element is counted as a single element regardless of whether it is a simple key or a complex datatype. StructCount() is unaware of anything beyond the level of the specified structure.

StructDelete()

The StructDelete() function deletes an element from a structure. It takes three arguments: the structure to be processed, the key to be deleted, and an optional boolean to specify whether a value should be returned if the key doesn't exist. The default for the third argument is FALSE, in which case the function always returns YES. If the third argument is TRUE, the function returns YES if the key exists in the structure, and NO if the key doesn't exist.

```
<cfset Students = StructNew()>
<cfset Students.firstName = "John">
<cfset Students.lastName = "Public">
<cfset Students.midInitial = "Q.">
Before:
<cfdump var="#Students#">
<cfset StructDelete(Students,"midInitial","YES")>
After:
<cfdump var="#Students#">
```

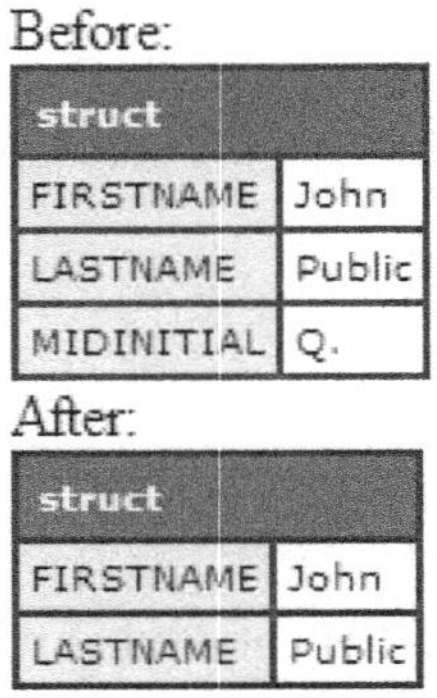

Illustration 31 - StructDelete()

StructFind()

The StructFind() function finds the value that corresponds to a specified key. It takes two arguments: the structure to be processed, and the key to have its value returned. The function returns the value if the key is found. If the first argument is of type other than a structure, or references a variable that doesn't exist, an error is thrown. If the second argument is not a valid key name, or the key does not exist, an error is thrown.

The real question regarding StructFind() is, "Why would I need to use this function?" My answer is, "You wouldn't." I have yet to find a structure use case where StructFind() is the only solution. Why does it exist? My best guess is just because ColdFusion's developers have always sought to make it very easy to use. Some users come from a programming background that makes them familiar with associative arrays (bracket notation); some come from an object-oriented background (dot notation), and other come from a background that emphasizes functions. No matter which background you're from, CFML keeps things comfortable.

StructFindKey()

The StructFindKey() function is the smarter big brother of the StructFind() function. It finds all the structures that contain the specified key in a structure of nested complex data. The base structure can contain other structures, arrays, and simple datatypes. The function takes three arguments: the structure to process, the key to find, and a string to indicate whether to return "ONE" or "ALL" of the results. StructFindKey() returns an array within which each matching structure is an element. This sounds very complex, but it's reasonably simple to understand with an example. The following code demonstrates how StructKeyFind() works, and the illustration shows the results.

```
<cfset Students = StructNew()>
<cfset Students.Student1.studentID = "12345">
<cfset Students.Student1.firstName = "John">
<cfset Students.Student1.lastName = "Public">
<cfset Students.HonorStudents = ArrayNew(1)>
<cfset Students.HonorStudents[1] = StructNew()>
<cfset Students.HonorStudents[1].studentID = "99999">
<cfset Students.HonorStudents[1].firstName = "Jane">
<cfset Students.HonorStudents[1].lastName = "Doe">
<cfset Students.HonorStudents[2] = StructNew()>
<cfset Students.HonorStudents[2].studentID = "44444">
<cfset Students.HonorStudents[2].firstName = "Albert">
<cfset Students.HonorStudents[2].lastName = "Onerock">
Structure to search:
<cfdump var="#Students#">
Results:
<cfset lastNames = StructFindKey(Students,"lastName","All")>
<cfdump var="#lastNames#">
```

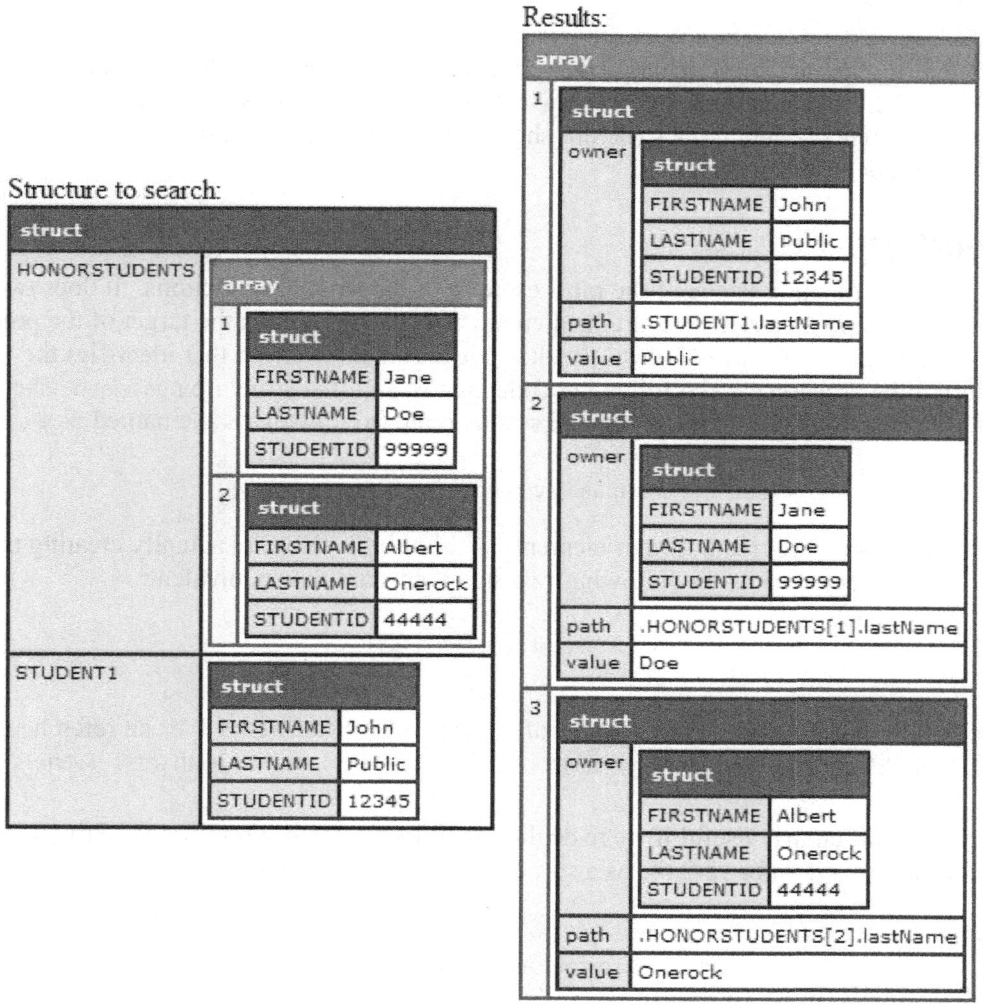

Illustration 32 - StructKeyFind()

Note how StructFindKey() basically homogenizes the references to all the found keys. We could now loop over the lastNames array and collect up each element's value key to get all the last names.

StructFindValue()

The StructFindValue() function works in exactly the same way as the StructFindKey() function, except that it searches for matching values instead of matching keys. In the example in Illustration 28, if we had used StructFindValue() to search for "Onerock", the result would be an array with one element – the structure shown in element 3 of the result in the StructKeyFind() illustration above.

StructGet()

The StructGet() function is one of the most esoteric of the structure functions. It does two things: create a pointer to a structure, and create the structure that is the target of the pointer if that structure doesn't already exist. It takes one argument: a string that identifies the structure to be processed. The following code creates a nested structure (Developmental inside Psychology inside Departments), and sets a pointer to it in a variable named devPsych:

```
<cfset devPsych = StructGet("Departments.Psychology.Developmental")>
```

This means if we now create further elements within devPsych, we're actually creating them in the larger structure, i.e., the following two statements are now equivalent:

```
<cfset Departments.Psychology.Developmental.Piaget = "PSYC200">
<cfset devPsych.Piaget = "PSYC200">
```

It also means that if we execute the latter of the above two statements, we can reference the result with #Departments.Psychology.Developmental.Piaget#, as well as with #devPsych.Piaget#.

This can be particularly useful if we're dealing with long structure references. StructGet() allows us to use a pointer variable as a sort of shorthand. For example:

```
<cfscript>
  emp = StructGet("Plant.Departments.Technical.Employees.Fred");
  emp.Salary = 47000;
  emp.Vacation = 160;
  emp.Lastname = "Flintstone";
</cfscript>
```

The result of this script is shown below.

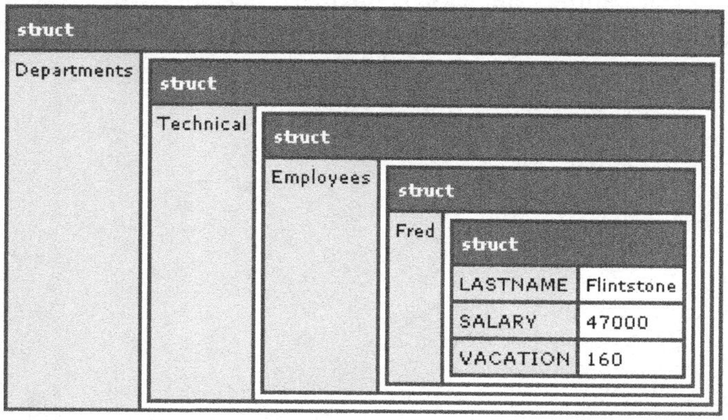

Illustration 33: StructGet() Example

StructInsert()

The StructInsert() function inserts a new element into a structure. It takes four arguments: the structure to process, the key of the element to add, the value of the element to add, and an optional boolean to specify whether overwrites are allowed. The last argument defaults to "NO". The function returns **YES** if the insertion is successful, and **NO** if an error occurs.

Basically, StructInsert() is the same as adding a new key with dot or bracket notation, with one very important difference. StructInsert() offers built-in protection against overwriting an existing key of the same name. If we try to do this, an error is thrown. With the fourth argument set to "YES", StructInsert() is functionally identical to using either of the direct notations.

StructIsEmpty()

The StructIsEmpty() function returns **YES** if the specified structure contains no data, and **NO** if the specified structure contains any data. It takes one argument: the structure to be processed. If the argument is of any type other than structure, or is omitted, an error is thrown.

StructKeyArray()

The StructKeyArray() function returns an array whose elements are the keys of the specified structure. It takes one argument: the structure to be processed. If the argument is of any type other than structure, or is omitted, an error is thrown. The following code demonstrates StructKeyArray(); its output is shown following.

```
<cfset Students = StructNew()>
<cfset Student.Student1 = StructNew()>
<cfset Students.Student1.studentID = "12345">
<cfset Students.Student1.firstName = "John">
<cfset Students.Student1.lastName = "Public">
<cfset Student.Student2 = StructNew()>
<cfset Students.Student2.studentID = "98765">
<cfset Students.Student2.firstName = "Jane">
<cfset Students.Student2.lastName = "Doe">
<cfset aryKeys = StructKeyArray(Students)>
<cfdump var="#aryKeys#">
```

array	
1	STUDENT1
2	STUDENT2

Illustration 34 - StructKeyArray()

Note that the contents of the array are only the keys of the Students array – NOT the entire structures contained as values of those keys. StructKeyArray() returns an array of keys only. If StructKeyArray() is used to process an empty structure, it returns an empty array.

StructKeyExists()

The StructKeyExists() function determines whether the specified key exists within the specified structure. It takes two arguments: the structure to search, and the name of the key to find. If either argument is omitted, or the first argument is not a structure, or the second argument is not a string, an error will be thrown. The function returns **YES** if the key is found and **NO** if it is not found. The function only checks the specified structure; it does not recurse into nested structures in search of the key. The following code demonstrates:

```
<cfset Students.Student1.firstName = "Dan">
<cfoutput>#StructKeyExists(Students,"Student1")> <!--- Returns YES --->
<cfoutput>#StructKeyExists(Students,"firstName")> <!--- Returns NO --->
```

If you need to find out whether a key exists anywhere within a nested structure, try using the StructFindKey() function.

StructKeyList()

The StructKeyList() function is the equivalent of the StructKeyArray() function, except that it returns a list instead of an array. It also takes an optional second argument: the delimiter to use for the list (defaults to comma). The following code repeats the example from the StructKeyArray() section, changing it to use StructKeyList().

```
<cfset Students = StructNew()>
<cfset Student.Student1 = StructNew()>
<cfset Students.Student1.studentID = "12345">
<cfset Students.Student1.firstName = "John">
<cfset Students.Student1.lastName = "Public">
<cfset Student.Student2 = StructNew()>
<cfset Students.Student2.studentID = "98765">
<cfset Students.Student2.firstName = "Jane">
<cfset Students.Student2.lastName = "Doe">
<cfset lstKeys = StructKeyList(Students)>
<cfdump var="#lstKeys#"> <!--- Outputs "STUDENT1,STUDENT2" --->
```

StructNew()

The StructNew() function creates a new empty structure. It takes no arguments, and returns a pointer to the new structure. You can see it in use throughout the sample code in this chapter.

PITFALL: The following code is valid, but useless:

```
<cfset StructNew()>
```

It creates a new empty structure without storing a pointer to it in a variable. The result is a structure in server memory that cannot be destroyed. Do this in a large loop and your server's memory will be impacted. Fortunately, CFMX's garbage collection seems to limit the impact to a maximum of about 32MB of RAM, so it won't be fatal to the server. (It's not so silly as it sounds; I came across this in a CF module in a production application once.)

StructSort()

The StructSort() function sorts the keys of the specified structure in terms of their corresponding values. It takes four arguments: the structure to be processed, an optional sort type (default is "TEXT"), an optional sort order (default is "ASC" for ascending; use "DESC" for descending), and an optional path to a subelement (to be used if sorting a nested structure). The function returns an array containing the sorted keys. The following code illustrates; the output follows.

```
<cfset Scores = StructNew()>
<cfscript>
 for (i = 1; i LT 7;i = i + 1) {
   StudentID = "Student#i#";
   Scores[StudentID] = Rand() * 100;
 }
 studentsByScore = StructSort(Scores,"numeric","DESC");
</cfscript>
Scores:
<cfdump var="#Scores#">
studentsByScore:
<cfdump var="#studentsByScore#">
```

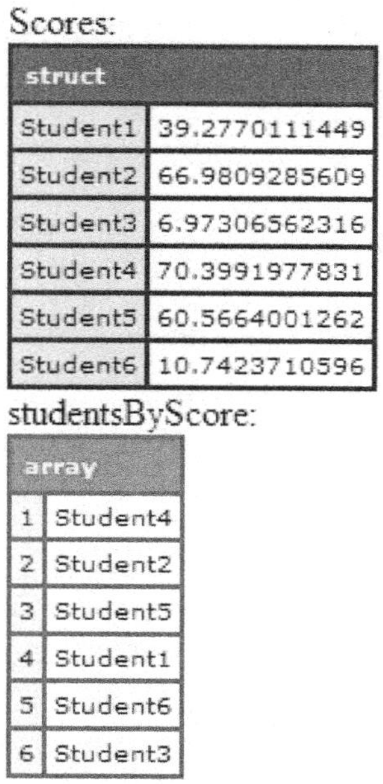

Scores:

struct	
Student1	39.2770111449
Student2	66.9809285609
Student3	6.97306562316
Student4	70.3991977831
Student5	60.5664001262
Student6	10.7423710596

studentsByScore:

array	
1	Student4
2	Student2
3	Student5
4	Student1
5	Student6
6	Student3

Illustration 35 - StructSort()

Note that when StructSort() operates, it does not return any values – just the keys sorted according to the values.

StructUpdate()

StructUpdate() sets the value for the specified key in the specified structure to the specified value. It takes three arguments: the structure to be processed, the key to reference, and the new value to assign to the key. If the first argument is of type other than structure, an error is thrown. If the second argument is of type other than string or is an invalid key name, an error is thrown. If the third argument is omitted, an error is thrown. If the specified key doesn't exist in the specified structure, an error is thrown. The function returns **YES** if the update is successful.

StructUpdate() differs from using the dotted notation or the bracketed notation in one important way. Using either of the direct notations, we can add or update a key/value pair with the same syntax. If the key doesn't exist, CF creates it. If we try to update a key that doesn't exist using StructUpdate(), however, an error is thrown. So if we have a condition where we need to handle the case of a nonexistent key prior to adding it, StructUpdate() is a good way to go. The following code demonstrates; the output follows:

```
<cfset Scores = StructNew()>
<cfscript>
 for (i = 1; i LT 7;i = i + 1) {
   StudentID = "Student#i#";
   Scores[StudentID] = Rand() * 100;
 }
</cfscript>
<!--- Simulate an external process that uses this structure --->
<cfset alertMsg = "">
<cfset strNewStudent = "Student7">
<cfset numNewScore = 100>
<cftry>
 <cfset temp = StructUpdate(Scores,strNewStudent,numNewScore)>
 <cfcatch type="Any">
   <cfset alertMsg = "#strNewStudent# added with a score of #numNewScore#">
   <cfset Scores[strNewStudent] = numNewScore>
 </cfcatch>
</cftry>
<cfif alertMsg GT "">
 <cfoutput>
 <script language="Javascript">
   alert('#alertMsg#');
 </script>
 </cfoutput>
</cfif>
<cfdump var="#Scores#">
```

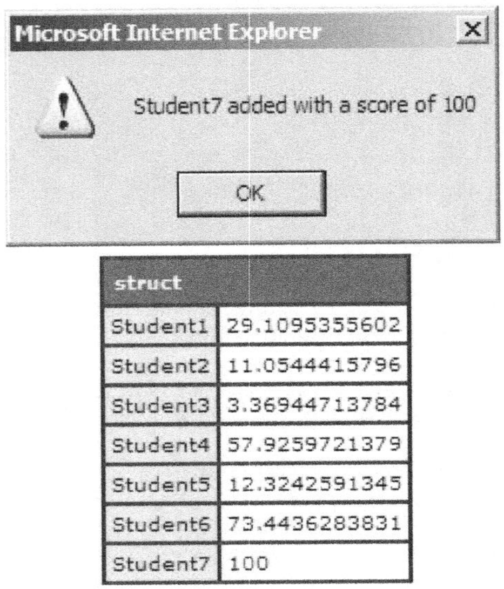

Illustration 36 - StructUpdate()

In the example, the Javascript alert is run as a result of the exception thrown by StructUpdate(), and the CFCATCH block handles the exception, sets the alert message and updates the structure using bracketed notation.

So that's our tour of the ColdFusion structure functions. Some are a bit esoteric, while you'll find others to be essential pieces of your daily CF arsenal. Armed with these powerful tools, we can make use of structures wherever they may benefit our work.

To further leverage our use of structures, we'll continue with an examination of the most common structures in ColdFusion programming – the ColdFusion variable scopes.

ColdFusion Scopes Are Structures

Many CF developers do not become aware of the nature of variable scopes until fairly well along in their ColdFusion careers. The earlier you get to know the nature of each scope, the better prepared you'll be to make programming decisions.

In ColdFusion MX and later, all of the following variable scopes are implemented as structures:

- Form
- URL
- Attributes
- Client
- Request
- Application
- Session
- Server
- CGI

We'll take a look at each of these scopes with a brief discussion and a look at how we can take advantage of the scope through structure manipulation. The basis of examination for this entire section is the following piece of code:

```
<!--- scopeStructs.cfm --->
<cfset request.foo = "bar">
<cfset client.foo = "bar">
<cfset application.foo = "bar">
<cfset session.foo = "bar">
<cfset server.foo = "bar">
<cfparam name="form.foo" default="bar">
<cfparam name="attributes.foo" default="bar">

<cfset lstScopes="Form,URL,Attributes,Client,Request,Application,Session,Server,CGI">

<cfloop list="#lstScopes#" index="thisScope">
  <cfif isDefined(thisScope)>
    <cfoutput>
      IsStruct(#thisScope#): #IsStruct(Evaluate(thisScope))#<br>
    </cfoutput>
    <cfdump var="#Evaluate(thisScope)#"><br>
  </cfif>
</cfloop>
```

We'll look at the output for each scope in its particular section. First off is the ever-popular FORM scope. (This discussion focuses on CFMX, but you can run the sample code above on your server to determine scope behaviors in whatever CFML application server you might be running, be it CF, BlueDragon, Railo, Smith Project, etc.)

Form Scope

The FORM scope exists regardless of whether the template was called from a form, but in order to properly look at it, we need to call our test code from a form. The following code works well for our purposes, and the illustration shows the relevant portion of the output.

```
<!--- formScopeStructs.cfm --->
<form action="scopeStructs.cfm" method="post">
 Foo: <input name="foo" type="Text">
 <input type="Submit" name="btnSubmit" value="submitted">
</form>
```

The FORM scope's structure includes a key for each input on the form, and the value passed for the input is associated with its key. In addition to the inputs specified on the form, ColdFusion adds the FIELDNAMES key. The value for this key is a list of the inputs

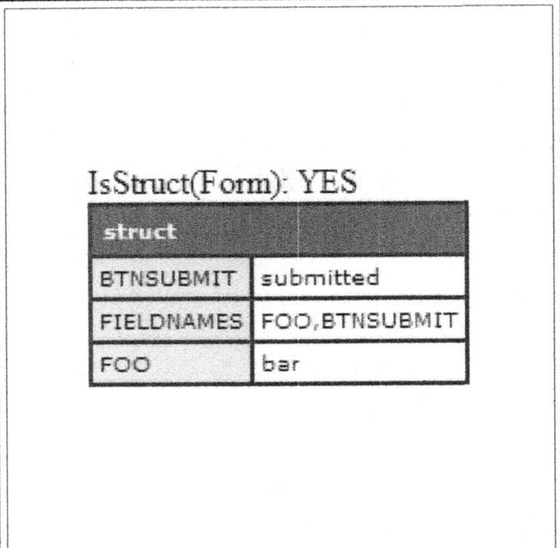

Illustration 37 - FORM Scope

on the form. In versions of CF before the FORM scope was presented as a structure, the FIELDNAMES list was the primary means of creating loops to look at, in turn, each value submitted from a form, as shown in the code below.

```
<cfloop list="#form.fieldnames#" index="thisField">
 <cfoutput>
  The value for the #thisField# field is #Evaluate("form." & thisField)#.<br>
 </cfoutput>
</cfloop>
```

Now that the scope itself is exposed as a structure, we can just loop over the structure to accomplish the same goal:

```
<cfloop collection="#form#" item="thisInput">
  <cfif thisInput NEQ "FIELDNAMES">
  <cfoutput>
   The value for the #thisInput# input is #form[thisInput]#.<br>
  </cfoutput>
  </cfif>
</cfloop>
```

Note that in order to get exactly the same output, we need to throw in the logic to ignore the FIELDNAMES field. This is not a general requirement; it's included here just to maintain the parallel examples.

URL Scope

Like the FORM scope, the URL scope is present regardless of whether it has any contents. In the case of the URL scope, there will only be keys in the structure if variables have been passed in the URL to call the template. For our purposes, we can call the test code with the following URL typed in the browser:

```
http://127.0.0.1/test/scopeStructs.cfm?foo=bar
```

The output of the relevant portion of the test code is shown below.

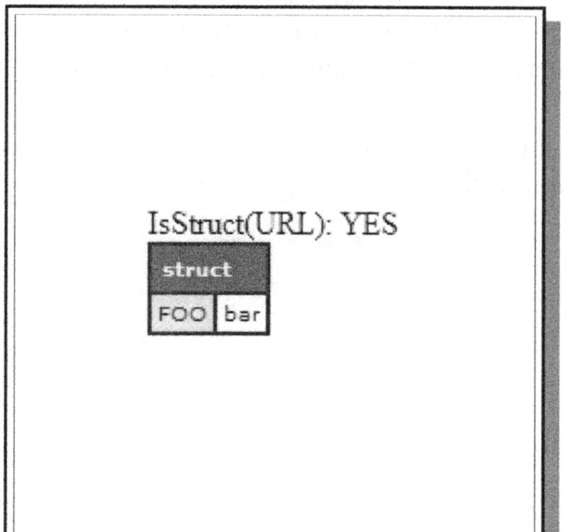

Illustration 38 - URL Scope

If no variables were passed to the template via the URL, the URL scope will be present and empty. That is, IsStruct(URL) and IsEmpty(URL) will both return TRUE.

Attributes Scope

Unlike the FORM and URL scopes, the ATTRIBUTES scope is not always present. It only exists if the template was called as a custom tag, either through CF_ syntax or using the <CFMODULE> tag. If we need to loop over the ATTRIBUTES scope, we need to make sure it exists first:

```
<cfif isDefined("attributes")>
  <!--- Attributes looping code here --->
</cfif>
```

NOTE: A major exception to this rule is Fusebox. Since Fusebox automatically copies all FORM and URL variables to ATTRIBUTES, the ATTRIBUTES scope always exists in a Fusebox application.

Client Scope

In CFMX, the CLIENT scope always exists, even if you don't define client variable use with the CLIENTMANAGEMENT attribute in a <CFAPPLICATION> tag. This can be useful, since we can test for client management by checking the value of IsEmpty(client). If there are variables present in the CLIENT scope, client management has been enabled. It's easy to forget that there might be an Application.cfm file in the directory structure above the current application that affects client management -- just something to keep in mind.

The CLIENT scope offers some very useful information about the current user's visit; the illustration shows the dump of the CLIENT scope structure from the sample code above.

IsStruct(Client): YES

struct	
cfid	200
cftoken	38612497
foo	bar
hitcount	61
lastvisit	{ts '2003-01-05 21:00:36'}
timecreated	{ts '2003-01-03 21:30:50'}
urltoken	CFID=200&CFTOKEN=38612497

Illustration 39 - CLIENT Scope

The variables CFID, CFTOKEN, HITCOUNT, LASTVISIT, TIMECREATED, and URLTOKEN always exist in the CLIENT scope.

Request Scope

The REQUEST scope structure is another one that always exists, regardless of whether we have created any variables within it. This structure is available to any template within the current HTTP request, including custom tags. Variables must be specifically assigned to the REQUEST scope.

Caller Scope

The CALLER scope is a bit of a strange duck. It refers to the memory space of the template that calls the current template, either through CF_ syntax or use of the <CFMODULE> tag. Consequently, the structure that holds CALLER-scoped variables does not exist unless the current tag has been called using one of these two methods. As with the ATTRIBUTES scope, we should test before trying to perform operations on the CALLER scope.

Some developers think of CALLER as the same as the calling page's VARIABLES scope, but this is not entirely accurate. We can use CALLER to access other scopes from the calling page as well, even though those other scopes are not exposed when we dump the CALLER structure. For example, consider the following two files:

```
<!--- ScopesByURL.cfm --->
URL Scope from calling page:
<cfdump var="#URL#">
<cfset variables.sitrep="fubar">
<cf_scopeStructs>
```

```
 <!--- scopeStructs.cfm --->
CALLER Scope:
<cfdump var="#caller#">
CALLER.URL:
<cfdump var="#caller.URL#">
```

ScopesByURL.cfm shows us the contents of the URL scope structure, then sets a variable called variables.sitrep and calls scopeStructs.cfm with a custom tag call. When called, scopeStructs.cfm dumps the CALLER scope structure and the URL scope structure from the calling page (referenced as caller.URL). For test purposes, we'll call ScopesByURL.cfm with this URL:

```
http://127.0.0.1/test/scopesByURL.cfm?sitrep=OK
```

The result is shown here:

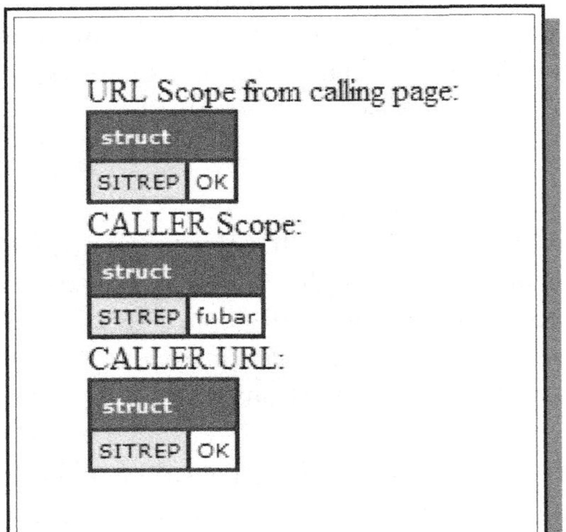

Illustration 40 - CALLER Scope

Note that even though the URL scope structure isn't exposed in the CALLER scope structure, we can still read the contents of CALLER.URL. So the CALLER scope structure only contains the variables defined in the calling template's VARIABLES scope, even though we can use the CALLER. prefix to reference the scope structures from the calling template.

Application Scope

The APPLICATION scope structure does not exist unless one of two thing occurs: the <CFAPPLICATION> tag is executed to define an application, or a variable is explicitly defined with the APPLICATION scope prefix. If an application has been defined, the variable APPLICATION.APPLICATIONNAME will exist in addition to any variables that have been defined within the scope.

Session Scope

The SESSION scope is very similar in behavior to the CLIENT scope, with a couple of important differences.

First, the SESSION scope structure does not contain the same set of default variables as the CLIENT scope structure. The next illustration shows the SESSION scope from our sample code.

The CFID, CFTOKEN, and URLTOKEN variables are the same as their counterparts in the CLIENT scope. The HITCOUNT, TIMECREATED, and LASTVISIT variables are not present in the SESSION scope. The SESSIONID variable, which is not in the CLIENT scope, is always present in the SESSION scope.

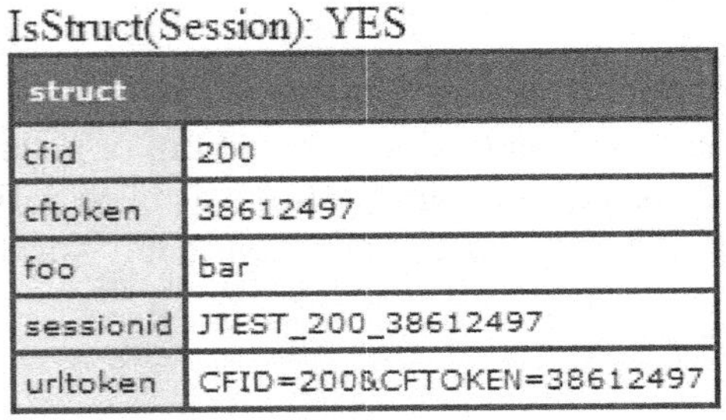

IsStruct(Session): YES

struct	
cfid	200
cftoken	38612497
foo	bar
sessionid	JTEST_200_38612497
urltoken	CFID=200&CFTOKEN=38612497

Illustration 41 - SESSION Scope

Second, the SESSION scope is server-specific. This means that if an application is implemented in a clustered server environment, a variable created in the SESSION scope will be unavailable if the cluster management software shifts the user to another server as part of its regular load balancing chores. The CLIENT scope is not server-specific, so it does not have this problem. (Some clustered configurations only direct entire sessions to a server, then leave the user there throughout their visit. If this is your configuration, you don't need to worry about losing session variables.)

Server Scope

The SERVER scope structure is always present, and contains a wealth of information about the server on which the application is running. The illustration is a dump of the server scope structure from our sample code.

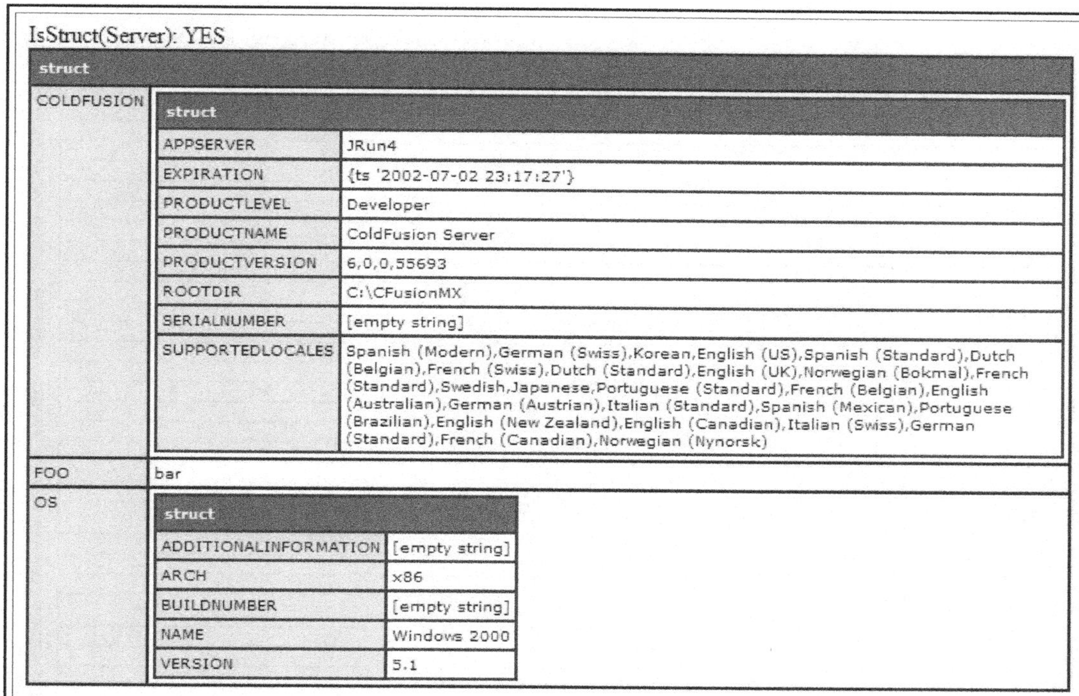

IsStruct(Server): YES

struct

COLDFUSION	APPSERVER	JRun4
	EXPIRATION	{ts '2002-07-02 23:17:27'}
	PRODUCTLEVEL	Developer
	PRODUCTNAME	ColdFusion Server
	PRODUCTVERSION	6,0,0,55693
	ROOTDIR	C:\CFusionMX
	SERIALNUMBER	[empty string]
	SUPPORTEDLOCALES	Spanish (Modern),German (Swiss),Korean,English (US),Spanish (Standard),Dutch (Belgian),French (Swiss),Dutch (Standard),English (UK),Norwegian (Bokmal),French (Standard),Swedish,Japanese,Portuguese (Standard),French (Belgian),English (Australian),German (Austrian),Italian (Standard),Spanish (Mexican),Portuguese (Brazilian),English (New Zealand),English (Canadian),Italian (Swiss),German (Standard),French (Canadian),Norwegian (Nynorsk)
FOO	bar	
OS	ADDITIONALINFORMATION	[empty string]
	ARCH	x86
	BUILDNUMBER	[empty string]
	NAME	Windows 2000
	VERSION	5.1

Illustration 42 - SERVER Scope

In addition to the variable we created, there are two structures contained within the SERVER scope structure: COLDFUSION and OS. (Note the interesting claim by CFMX 6 that I'm running Windows 2000 Version 5.1. That's actually Windows XP Professional. This is corrected in CFMX 7.)

The variables in the SERVER scope can be particularly useful when deploying on a shared hosting server. We may not have access to a lot of information about the server; the SERVER scope structure is a good starting point to find out what resources we have.

CGI Scope

Like the SERVER scope, the CGI scope structure contains a wealth of information about the current HTTP request. The illustration shows this information as produced by the example code.

IsStruct(CGI): YES

struct	
AUTH_PASSWORD	[empty string]
AUTH_TYPE	[empty string]
AUTH_USER	[empty string]
CERT_COOKIE	[empty string]
CERT_FLAGS	[empty string]
CERT_ISSUER	[empty string]
CERT_KEYSIZE	[empty string]
CERT_SECRETKEYSIZE	[empty string]
CERT_SERIALNUMBER	[empty string]
CERT_SERVER_ISSUER	[empty string]
CERT_SERVER_SUBJECT	[empty string]
CERT_SUBJECT	[empty string]
CF_TEMPLATE_PATH	C:\Data\webroot\test\scopeStructs.cfm
CONTENT_LENGTH	[empty string]
CONTENT_TYPE	[empty string]
CONTEXT_PATH	[empty string]
GATEWAY_INTERFACE	CGI/1.1
HTTPS	[empty string]
HTTPS_KEYSIZE	[empty string]
HTTPS_SECRETKEYSIZE	[empty string]
HTTPS_SERVER_ISSUER	[empty string]
HTTPS_SERVER_SUBJECT	[empty string]
HTTP_ACCEPT	image/gif, image/x-xbitmap, image/jpeg, image/pjpeg, application/vnd.ms-powerpoint, application/vnd.ms-excel, application/msword, application/x-shockwave-flash, */*
HTTP_ACCEPT_ENCODING	gzip, deflate
HTTP_ACCEPT_LANGUAGE	en-us
HTTP_CONNECTION	Keep-Alive
HTTP_COOKIE	cf_debug_variables=block; cf_debug_application=none; cf_debug_cgi=none; cf_debug_server=none; cf_debug_session=none; cf_debug_general=none; cf_debug_request=none; HIDEWELCOMEWINDOW=1; CFID=200; CFTOKEN=38612497
HTTP_HOST	127.0.0.1
HTTP_REFERER	[empty string]
HTTP_USER_AGENT	Mozilla/4.0 (compatible; MSIE 6.0; Windows NT 5.1; .NET CLR 1.0.3705)
PATH_INFO	[empty string]
PATH_TRANSLATED	C:\Data\webroot\test\scopeStructs.cfm
QUERY_STRING	[empty string]
REMOTE_ADDR	127.0.0.1
REMOTE_HOST	127.0.0.1
REMOTE_USER	[empty string]
REQUEST_METHOD	GET
SCRIPT_NAME	/test/scopeStructs.cfm
SERVER_NAME	127.0.0.1
SERVER_PORT	80
SERVER_PORT_SECURE	0
SERVER_PROTOCOL	HTTP/1.1
SERVER_SOFTWARE	Apache/1.3.27 (Win32) PHP/4.2.3
WEB_SERVER_API	[empty string]

Illustration 43 - CGI Scope

A complete discussion of the CGI scope structure and its many keys is beyond the scope of this book, but you can see from the example above (and if you run the example code on your own machine) that this scope can be an invaluable source of information.

Having visited all the ColdFusion structure manipulation functions and discussed all the scope structures, we've covered the subject of structures fairly thoroughly. At this point, you know more about structures than perhaps 90% of ColdFusion developers. We'll cap off the chapter by looking at a few practical examples of structures in action.

Structure Examples

Structures can be used in a practically infinite number of situations to solve common problems. Besides being able to use structure manipulation on the ColdFusion scope structures, we can create our own structures to solve a variety of problems. The following sections provide a few common problems and simple solutions for them.

Structure Example 1: State Names

A structure to hold state abbreviations and names is a good starting example for structure manipulation. You can copy this code into any application where you take abbreviations as input and need to produce the full name for output. (This example is admittedly slanted to American readers; feel free to substitute abbreviations and names for the political divisions of your favorite country.)

```
<cfscript>
  States = StructNew();  <!--- This line is optional in CFMX and later --->
  States.AL="Alabama";
  States.AK="Alaska";
<!-- Other states skipped to save space --->
  States.VA="Virginia";
  States.WA="Washington";
  States.WV="West Virginia";
  States.WI="Wisconsin";
  States.WY="Wyoming";
</cfscript>
```

With this structure defined, translating abbreviations is simple:

```
<cfset myState="VA">
<cfoutput>#States[myState]#</cfoutput>
```

But what if we want to get a little more sophisticated? Maybe we'd like to be able to return the state's name and its capital. In this case, we'd just change the definitions a little bit:

```
<cfscript>
 States = StructNew();
 States.AL=StructNew();
 States.AL.Name="Alabama";
 States.AL.Capital="Montgomery";
<!-- Other states skipped to save space --->
</cfscript>
```

or, if we're using ColdFusion 8:

```
<cfscript>
 States.AL={Name="Alabama", Capital="Montgomery"};
 // Other states skipped to save space
</cfscript>
```

Now we can get a state's name and capital by referencing (for Alabama as an example) `States.AL.Name` and `States.AL.Capital`, respectively.

Of course, the values for this structure shouldn't be hardcoded, so we should probably have a bit of code to pull them from a database table:

```
<cfquery name="qryGetStates" datasource="myData">
 SELECT stateAbbrev,
        stateName,
        capital
   FROM States
</cfquery>

<cfset States = StructNew()>
<cfloop query="qryGetStates">
 <cfscript>
 States[stateAbbrev]=StructNew();
 States[stateAbbrev].Name="#stateName#";
 States[stateAbbrev].Capital="#capital#";
 </cfscript>
</cfloop>
```

And lastly, you may be wondering why we'd bother doing this anyway. After all, if the recordset is in memory, why translate it into a structure? The answer is that a structure makes referencing the values much easier. Let's see what we need to do to get the name and capital for a state from a recordset named `qryGetStates` in memory:

```
<cfset myStateAbbrev="AL">
<cfquery name="qryGetMyState" dbtype="query">
 SELECT Name,
       Capital
  FROM qryGetStates
 WHERE stateAbbrev = '#myStateAbbrev#'
</cfquery>

<cfoutput>
My state is #qryGetMyState.Name#.<br>
Its capital is #qryGetMyState.Capital#.
</cfoutput>
```

Contrast the above code with what's needed to retrieve the same data from the nested structure:

```
<cfset myStateAbbrev="AL">
<cfoutput>
My state is #States[myStateAbbrev].Name#.<br>
Its capital is #States[myStateAbbrev].Capital#.
</cfoutput>
```

So if our application does a lot of work with state abbreviations, names, and capitals, it would be a good idea to create the structure in a persistent variable and use it instead of manipulating recordsets all the time. Since the values don't change frequently, we could store the structure as an application-scope variable and it would always be available.

Speaking of structures that should always be available, let's take a look at our next example: the ever-popular shopping cart.

Structure Example 2: Shopping Cart

Most web developers are presented with the task of managing a shopping cart at some point in their careers. Many different approaches have been taken to managing shopping cart data within applications, and many of them are very convoluted. Hopefully this example will give you some good ideas about how to build a straightforward cart.

One of the keys to modeling something in a program is to make it operate as much like its real-world counterpart as possible. In other words, build a model.

So let's consider a shopping cart. It has some important characteristics. First, it belongs to me. That is, as long as I'm at the store, I push the same cart around. Second, I can add items to the cart and take them out at random. I can add more of an item if I want, at random. I can put some of an item back out of the cart if I want, at random. That's about all the cart needs, so let's think about items.

An item, or product, has some characteristics as well. I can see how many of it are in the cart. I can tell how much each one costs. And I can identify each item as being different from each of the other items. Just for fun, let's add another characteristic: whether the item is a "bonus" item in our store.

So let's build a cart. It sounds like it should be some kind of container – a variable that can hold other variables. At first, it might seem like an array would be a good solution, but we're going to use a structure, for reasons to be disclosed later. So let's create a structure:

```
<cfset variables.cart = StructNew()>
```

This is a good start. We have a variable that can hold other stuff. Now let's look at how we'll store items. Each item has some information we'll need to be able to look up easily: price, quantity, and productID. This is definitely a recipe for a structure.

So when we need to add an item to the cart, we just need to define a new item structure within the cart structure:

```
<cfset variables.cart.AppleButter = StructNew()>
```

For better flexibility, we'll probably want to use a slightly different syntax:

```
<cfset productToAdd = "AppleButter">
<cfset variables.cart[productToAdd] = StructNew()>
```

Then we can specify the quantity of that item to put in the cart (we won't worry about price right now – that doesn't matter until checkout) and whether it's a bonus item for this week:

```
<cfset variables.cart.AppleButter.qty = 3>
<cfset variables.cart.AppleButter.bonus = false>
```

Incidentally, if you're using CFMX, the whole process of creating the cart and dropping three jars of apple butter in it can be condensed into the last two statements. CFMX handles the creation of the structures without a hiccup. Further, if you're using ColdFusion 8, you can condense the whole thing into the single statement `<cfset variables.cart.AppleButter = {qty=3, bonus=false}>`. Whether you use the CFMX or ColdFusion 8 shortcuts or the full set of CFSETs for a previous version of CF, the result is the nested structure shown here:

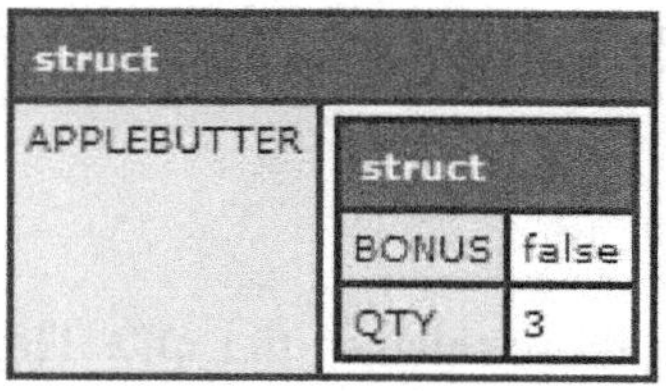

Illustration 44 - Cart Structure

We now have a very easy way to add items to the cart – just add a new element to the structure, along with the quantity and whether it's a bonus item. Let's add a box of raisins:

```
<cfset variables.cart.Raisins.qty = 1>
<cfset variables.cart.Raisins.bonus = true>
```

Now our cart has a couple of items, as seen here:

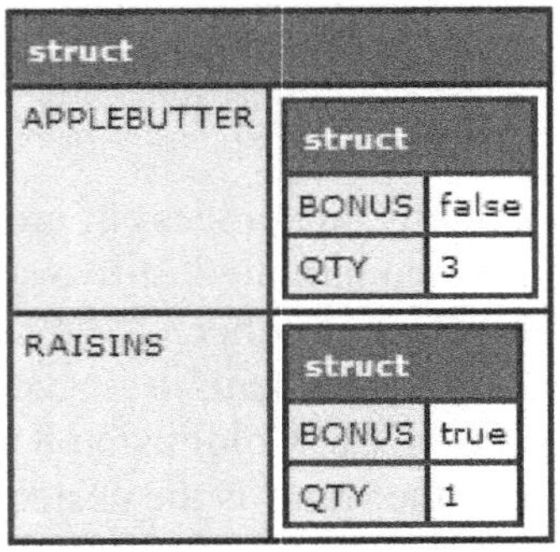

Illustration 45 - Added Item

Now what happens if I decide I only want one jar of apple butter? We just set the quantity to the appropriate value:

```
<cfset variables.cart.AppleButter.qty = 1>
```

And what if I decide I didn't want those raisins after all? This is why we chose to use a structure for my cart instead of an array. If we were using an array, we would need to determine which element of the array contains the item we want to delete before we could delete it. With our structure arrangement, a simple `StructDelete()` function does the trick:

```
<cfset temp = StructDelete(variables.cart,"Raisins")>
```

And now our cart is back to a single item, as shown below.

Now that we can successfully create a cart, add items to it, change quantities, and take items out, there's only one more characteristic left to address: the cart needs to belong to me. This is a bit of a problem in ColdFusion. The natural inclination is to create the cart structure as a client variable. Unfortunately, complex datatypes (like structures) can't be stored in client variables. We could store the cart in a session variable. This would be fine so long as the application doesn't live in a clustered server environment. Session variables do not persist across clustered servers.

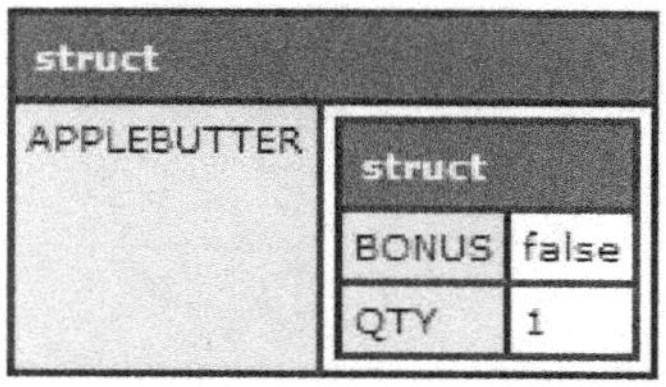

Illustration 46 - Deleted Item

Let's assume we want to use a client variable. So we need a way to convert our structure into a string. This is just the sort of job that WDDX was designed to handle. The <CFWDDX> tag can convert the structure into a string so we can save it as a client variable. The syntax looks like this:

```
<cfwddx action="CFML2WDDX" input="#variables.cart#" output="client.cart">
```

And the result is the variable client.cart, which holds a string that looks like this (carriage returns and indenting added for readability):

```
<wddxPacket version='1.0'><header/>
<data>
<struct>
 <var name='APPLEBUTTER'>
  <struct>
   <var name='BONUS'>
    <string>false</string>
   </var>
   <var name='QTY'>
    <string>1</string>
   </var>
  </struct>
 </var>
</struct>
</data>
</wddxPacket>
```

You can see that the structures' data are preserved, as well as the format, ready to be reconstituted into a new nested structure when the cart is needed next. This is also done using the <CFWDDX> tag:

```
<cfwddx action="WDDX2CFML" input="#client.cart#" output="variables.cart">
```

The general model is simple: in any template where we need to manipulate the cart, we start by using <CFWDDX> to convert client.cart into variables.cart, then do whatever processing is required, and finally use <CFWDDX> to convert variables.cart back into client.cart. A more complete discussion of WDDX can be found in the next chapter.

Summary

Structures are among the most useful complex data types. ColdFusion's version of associative arrays are easy to manipulate, and yet very powerful in their ability to store and assist in the management of very complex combinations of data.

This chapter is the last in our direct discussion of ColdFusion complex data types. The next chapter explores a related topic – Web Distributed Data eXchange (WDDX).

4.WDDX and Javascript

When taking advantage of complex data types, there are times when we come up against places where we'd like to use them, but there are hurdles in the way. For example, some scopes in ColdFusion don't accept complex data types (the `client` scope is one). Another common hurdle is the desire to move data to and from Javascript, to allow sophisticated front-end manipulation. Both of these examples can be managed with relative ease by using WDDX.

What is WDDX?

Web Distributed Data eXchange (WDDX) was originally designed by Allaire. As its name implies, WDDX is intended to be a standardized means of exchanging data between environments that might otherwise be isolated from one another. The definition of WDDX, as provided by OpenWDDX.org, states that

> *WDDX is an XML-based technology that enables the exchange of complex data between Web programming languages, creating what some refer to as 'Web syndicate networks'. WDDX consists of a language-independent representation of data based on an XML 1.0 DTD, and a set of modules for a wide variety of languages that use WDDX. WDDX can be used with HTTP, SMTP, POP, FTP and other Internet protocols that support transferring textual data.*

Essentially, WDDX provides a way to convert a complex data type into a simple string, encoded with the WDDX XML vocabulary. This conversion is known as serializing a WDDX packet. At the other end of the exchange, the WDDX packet is converted from a string into a complex data type that is understood by the receiving system. This conversion is known as deserializing a WDDX packet. The entire exchange is shown in this illustration:

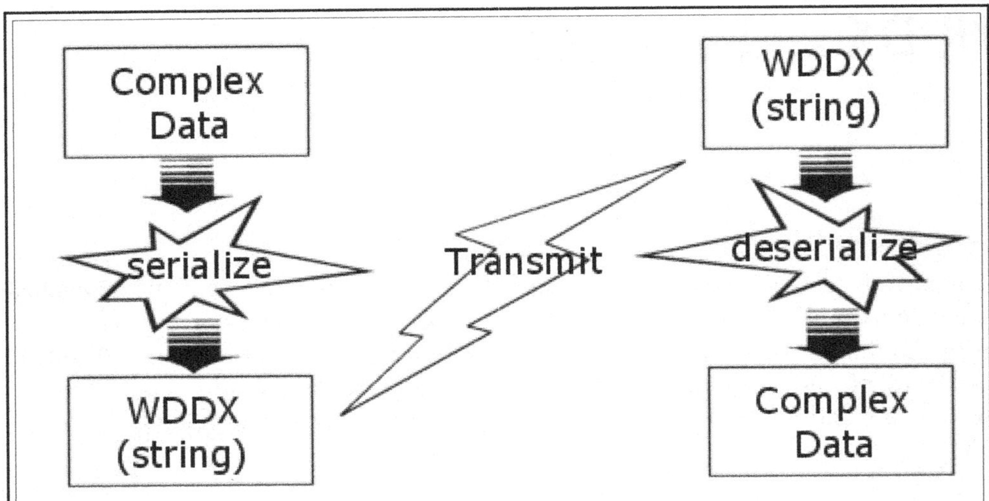

Illustration 47 - WDDX Exchange

Since strings can be transmitted between essentially any two systems or environments, the WDDX packet can be easily transferred from the originating system to the receiving system. What happens once the packet arrives at its destination is entirely up to the receiver.

Serializing and Deserializing

A WDDX packet is just an XML data set. As such, it can be processed by any system that has facilities to deal with XML. Even if a system doesn't have formal WDDX support, it is possible to create WDDX handling facilities on practically any platform. This is one of the features of WDDX that make it very appealing.

For our purposes, we only need to consider two environments when dealing with WDDX packets: ColdFusion and Javascript. There are times when both the sender and receiver will use ColdFusion, times when the sender will use ColdFusion and the receiver will use Javascript, and vice versa. CFML gives us a very simple way to handle all of these cases: the `<CFWDDX>` tag. This tag handles serializing and deserializing for both CF and Javascript. We'll work through some examples to illustrate how `<CFWDDX>` works.

Using Client Scope with Complex Data

The first case we'll look at is storing complex data in CF's client scope. We already know that CF won't store complex data types in the client scope. If we want to pass complex data from one CF-based request to another through the use of a client variable, we first need a way to convert the complex data to a simple string. This is a job for <CFWDDX>!

For this example, we'll use a structure named variables.cart. This is the shopping cart for the user. Its contents are shown here:

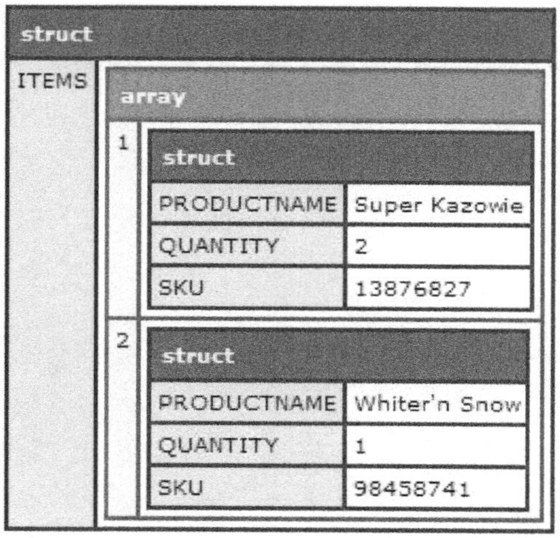

Illustration 48 - variables.cart

We want this cart variable to follow the user through their visit – a perfect use for a client-scoped variable. Since we can't store the cart directly as a client variable, we serialize it into a WDDX packet:

```
<cfwddx action="CFML2WDDX"
        input="#variables.cart#"
       output="client.cart">
```

The result of the serialization is the client.cart variable, which look like this (lines and indenting added for readability):

```
<wddxPacket version='1.0'>
 <header/>
 <data>
  <struct>
   <var name='ITEMS'>
    <array length='2'>
     <struct>
      <var name='SKU'>
       <string>13876827</string>
      </var>
      <var name='PRODUCTNAME'>
       <string>Super Kazowie</string>
       </var>
      <var name='QUANTITY'>
       <string>2</string>
      </var>
     </struct>
     <struct>
      <var name='SKU'>
       <string>98458741</string>
      </var>
      <var name='PRODUCTNAME'>
       <string>Whiter'n Snow</string>
      </var>
      <var name='QUANTITY'>
       <string>1</string>
      </var>
     </struct>
    </array>
   </var>
  </struct>
 </data>
</wddxPacket>
```

With the cart data safely stored in a client variable, we can make use of it in other templates and processes. When we want to use the data in the cart, we start by deserializing the WDDX packet back into its native CF form in the variables.cart variable:

```
<cfwddx action="WDDX2CFML"
        input="#client.cart#"
       output="variables.cart">
```

The data has been returned to its original state, and is ready to be manipulated by CF. If examined using <CFDUMP>, the output is identical to that in Illustration 47.

The above example demonstrates serializing data into WDDX for storage, then deserializing it for processing. The <CFWDDX> tag allows us to take a slightly different approach for the case of exchanging data between CF and Javascript.

Exchanging Data with Javascript

When <CFWDDX> was designed, its creators considered the special relationship between CFML and Javascript. These two technologies are often found working in close proximity to one another, CF on the server and Javascript on the browser. Prior to <CFWDDX>, the only way to achieve interaction between CF and Javascript was to use CF to write Javascript which was then executed by the browser. Unfortunately, there wasn't a way to communicate complex Javascript data back to CF.

ColdFusion to Javascript

When converting data from CF to Javascript, <CFWDDX> provides a way to do so directly, without storing the data in an intermediate WDDX packet. This makes sense, since CF is responsible for constructing the document that gets sent to the browser, including any Javascript included in it. It wouldn't make sense to create a WDDX packet and then send it, along with the Javascript code to deserialize it, to the browser.

So <CFWDDX> provides the CFML2JS action. This action simply converts the complex CF data directly into its corresponding Javascript data type. Using the example of our shopping cart again, we could use the following code:

```
<cfwddx action="CFML2JS" input="#variables.cart#" toplevelvariable="myCart">
```

This tag creates the Javascript code shown below.

```
myCart = new Object();
myCart["items"] =  new Array();
myCart["items"][0] = new Object();
myCart["items"][0]["sku"] = "13876827";
myCart["items"][0]["productname"] = "Super Kazowie";
myCart["items"][0]["quantity"] = "2";
myCart["items"][1] = new Object();
myCart["items"][1]["sku"] = "98458741";
myCart["items"][1]["productname"] = "Whiter'n Snow";
myCart["items"][1]["quantity"] = "1";
```

Remember that this is Javascript: the surrounding <script> and </script> tags are not shown in this example, but the <CFWDDX> tag must occur within a <script language="Javascript"> block because it generates Javascript code. The myCart associative array can now be used within the page's Javascript.

> **NOTE:** If a WDDX packet is received from another source, <CFWDDX> can make it available to Javascript by using the "WDDX2JS" value for the action attribute.

When using CFCs in ColdFusion 8, we can also use the returntype="WDDX" attribute to specify that the return from the function must be valid WDDX. This adds a layer of error-checking when using CFCs to generate WDDX data.

Javascript to ColdFusion

The trip to carry complex data from Javascript back to ColdFusion is slightly more involved, and requires the use of non-CF resources. Since Javascript executes on the browser, ColdFusion doesn't have a chance to interact with Javascript data unless that data is sent to the CF server. So we need a way to serialize complex data in Javascript and send it to ColdFusion. The wddx.js Javascript library provides the way, and we can download it from OpenWDDX.org.

To use the wddx.js library, its functions must be made available to the template. The following code takes care of it:

```
<script language="Javascript" src="wddx.js"></script>
```

After the library is loaded, we can use the wddxSerialize() function to turn Javascript data into a WDDX packet:

```
<script language="Javascript">
  mySerializer = new WddxSerializer;
  wddxPacket = mySerializer.serialize(aryRecords);
</script>
```

Finally, we need a way to send the WDDX packet to the CF server. We can do this by storing the WDDX packet as the value of a hidden form input and submitting the form:

```
<form name="myForm" action="processData.cfm" method="post">
  <input type="hidden" name="inWDDX">
</form>
<script language="Javascript">
  document.myForm.inWDDX.value = wddxPacket;
  document.myForm.submit();
</script>
```

Using this method, the WDDX packet gets submitted to the processData.cfm template as a variable named form.inWDDX. We can use <CFWDDX> to deserialize the WDDX packet for use in the template:

```
<CFWDDX action="WDDX2CF" input="#form.inWDDX#" output="aryFromJavascript">
```

This completes the Javascript-to-ColdFusion loop.

Summary

While complex data types such as arrays and structures are very powerful tools in ColdFusion, there are times when we need some assistance in moving complex data among various environments. The WDDX standard provides a way to convert complex data into structured string data, allowing it to be transferred among a broad variety of systems.

Two of the most frequent uses of WDDX in ColdFusion are the storing of complex data in client variables, and moving complex data to and from Javascript. The <CFWDDX> tag facilitates both of these tasks.

WDDX provides one means of moving data from the ColdFusion (server) side of things to the Javascript (browser) side. In similar fashion, we can also use Javascript Object Notation (JSON) to express CF-based data for use by Javascript. JSON is frequently used within AJAX-based applications. The next chapter covers JSON and its manipulation in ColdFusion.

5.JSON in ColdFusion

Javascript Object Notation (JSON) was designed to serve a similar purpose to WDDX. It is a notation that can be used to express complex datatypes in string format. This notation makes transporting complex data between systems and between tiers within systems possible. The rest of this chapter explores JSON and how it is implemented in ColdFusion 8.

What is JSON?

JSON is a data description language, also known as a data interchange format. This just means that JSON provides a way to describe, in plain text, a variety of complex data types. This renders the complex data as a simple string, which makes it easy to exchange the data within and among systems. The official JSON standard is codified in RFC 4627.

> **NOTE:** Where turning a complex datatype into text using WDDX is referred to as "serializing", the JSON community refers to the same process as "stringifying". Go figure. Adobe chose to stick to convention and use "serializing".

In the typical process for using JSON-encoded data, a server process creates the data, then stringifies (serializes) it into JSON. The JSON string is sent to the client, where it is destringified (deseriaized). The entire exchange is shown here:

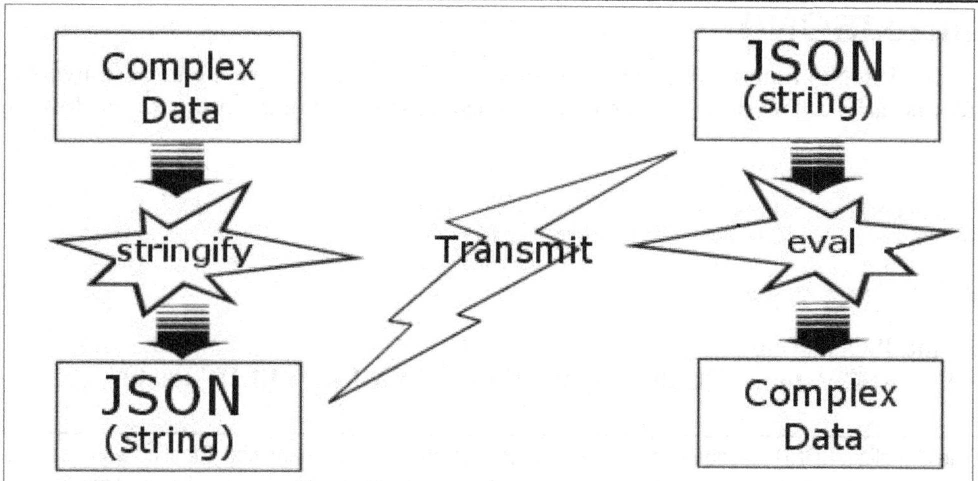

Illustration 49 - JSON Exchange

Since strings can be transmitted between essentially any two systems or environments, the JSON packet can be easily transferred from the originating system to the receiving system, which is typically a Javascript environment. What happens once the packet arrives at its destination is entirely up to the receiver. The examples later in this chapter will demonstrate a couple of different techniques.

Stringifying and Destringifying

JSON is actually a subset of the Javascript object literal notation. As such, a JSON string can be "destringified" by simply evaluating it within the Javascript environment. So when a Javascript receives a piece of JSON, it can reconstitute the complex data with a statement like this:

```
var complexData = eval(JSONdata);
```

For our purposes, we only need to consider two environments when dealing with JSON data: ColdFusion and Javascript. We will produce the JSON data using ColdFusion, and consume it using Javascript. There are variations on this theme, but the basics run true regardless.

ColdFusion 8 provides means to deal with JSON data. There are functions to stringify and destringify (which ColdFusion calls, I think reasonably, SerializeJSON() and DeserializeJSON(); an IsJSON() function; and an easy means to cause CFC methods to return data in JSON. We'll work through some examples to illustrate how each of these tools works.

SerializeJSON()

The SerializeJSON() function takes one argument: a ColdFusion variable to be converted into JSON notation. The output is the resulting JSON string. For example, the following code:

```
<cfset EmployeeStruct = {firstName="John", lastName="Doe", employeeID="000080"}>
<cfset EmployeeJSON = SerializeJSON(EmployeeStruct)>

<cfdump var="#EmployeeJSON#">
```

Produces the JSON string
{"FIRSTNAME":"John","LASTNAME":"Doe","EMPLOYEEID":80.0}

> **NOTE:** The SerializeJSON() function can also take an optional second argument when dealing with queries. Queries are outside the scope of this book; consult the ColdFusion 8 documentation for details regarding queries.

DeSerializeJSON()

The DeSerializeJSON() function also takes one argument: a JSON string to be converted into a ColdFusion datatype. The output is the resulting datatype. For example, the following code:

```
<cfset tempsArray = DeSerializeJSON("[65.0,87.0,66.0,85.0,88.0,92.0,95.0]")>
<cfdump var="#tempsArray#">
```

Produces the ColdFusion array shown here:

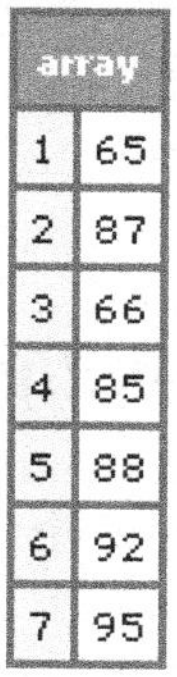

array	
1	65
2	87
3	66
4	85
5	88
6	92
7	95

Illustration 50: Array from JSON

IsJSON()

The IsJSON() function takes one argument: a value to be examined. If the value is a string that contains valid JSON notation, the function returns **YES**. If the value is anything else, the function returns **NO**. So if we examine the JSON string from the previous example, like this:

```
<cfset jsonCheck = IsJSON("[65.0,87.0,66.0,85.0,88.0,92.0,95.0]")>
```

The variable jsonCheck will be set to **YES**. On the other hand, if we introduce an error into the string, such as removing the closing bracket:

```
<cfset jsonCheck = IsJSON("[65.0,87.0,66.0,85.0,88.0,92.0,95.0")>
```

The variable jsonCheck will be set to **NO**. If the variable to be examined is another datatype, such as a plain numeric, list, array, or structure, the function will return **NO**.

CFC ReturnType="JSON"

When using ColdFusion 8, we can specify a returntype attribute with a value of "JSON" in any <CFFUNCTION> tag in order to specify that the returned value of the function must be valid JSON. This is very useful when creating methods for use in AJAX applications.

Exchanging Data with Javascript

Like <CFWDDX>, JSON can work in close formation with CFML and Javascript. Unlike WDDX, JSON does not need a library in order to use JSON data—we just evaluate the notation in order to get Javascript data.

ColdFusion to Javascript Directly

One way to produce Javascript complex data using ColdFusion is to depend on the fact that CFML is run first—on the server—and the resulting HTML is then run on the browser. So we can use CFML to dynamically build Javascript that creates data using JSON. This practice has been commonly used to create things like cascading selects. Here's an example that demonstrates the use of CFML to generate a Javascript array:

```
<cfscript>
  myArray = ArrayNew(1);
  ArrayAppend(myArray,"Able");
  ArrayAppend(myArray,"Baker");
  ArrayAppend(myArray,"Charlie");
  myArrayJSON = SerializeJSON(myArray);
</cfscript>

<cfoutput>
<script language="Javascript">
 peopleArray = eval(#myArrayJSON#);
 myVar = "";
 for (i=0;i < peopleArray.length;i++){
   myVar = myVar + peopleArray[i] + ' ';
 }
 alert(myVar);
</script>
</cfoutput>
```

When this CFML template is executed, it generates the following HTML:

```
<script language="Javascript">
 peopleArray = eval(["Able","Baker","Charlie"]);
 myVar = "";
 for (i=0;i < peopleArray.length;i++){
   myVar = myVar + peopleArray[i] + ' ';
 }
 alert(myVar);
</script>
```

Notice the JSON inside the eval() function on the second line. When the browser renders this code, it generates the alert box shown here:

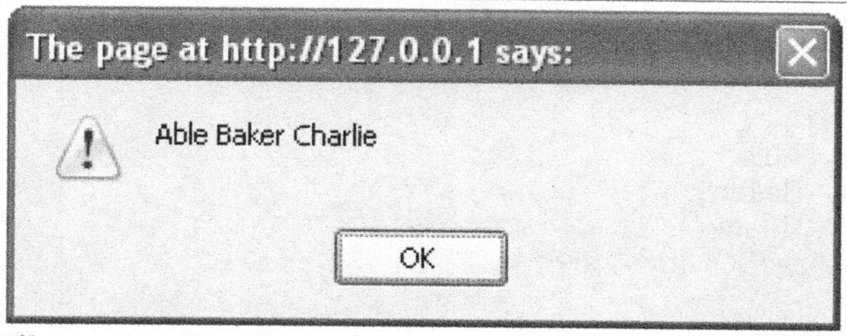

Illustration 51: JSON-generated Alert Box

This method of directly generating Javascript data using CFML is very handy when we know what data we need at design time. But it's not such a good solution when we want to dynamically fetch data into Javascript. For that we need a different approach, such as AJAX.

ColdFusion to Javascript Via AJAX

AJAX is a subject that needs much more space than we have here to discuss well. However, one of the basic techniques in an AJAX application is the use of the Javascript XmlHttpRequest object to retrieve data from a URL. In our case, we're talking about a URL for a CFML template that grabs some data and outputs it using JSON. We'll use the same data we used in the previous section, but instead of a CFML template that builds a Javascript, we'll use two templates: some HTML with a Javascript that calls a CFML template, and the CFML template that gets called. Note that this is NOT a complete code sample, as there would normally be some setting up of the XmlHttpRequest object, as well as dynamic URL specification, error handling, etc. I'm leaving these parts out here for the sake of brevity.

We'll start by looking at the CFML template that generates the JSON. It's very similar to the previous example, but note the addition of the <cfsetting enablecfoutputonly="yes"> tag—this ensures that there is no extra whitespace in our JSON output.

```
<!--- peopleArrayJSON.cfm --->
<cfsetting enablecfoutputonly="yes">
<cfscript>
  myArray = ArrayNew(1);
  ArrayAppend(myArray,"Able");
  ArrayAppend(myArray,"Baker");
  ArrayAppend(myArray,"Charlie");
  myArrayJSON = SerializeJSON(myArray);
</cfscript>
<cfoutput>#myArrayJSON#</cfoutput>
```

This just produces the same JSON array we saw before. Now we need the HTML template that will use an XmlHttpRequest object to call peopleArrayJSON.cfm:

```
<script language="Javascript">
var req = null;
if (window.XMLHttpRequest){
  req=new XMLHttpRequest();
} else if (window.ActiveXObject){
  req=new ActiveXObject('Microsoft.XMLHTTP');
}

if (req){
  req.onreadystatechange=showArray;
  req.open('GET','peopleArrayJSON.cfm',true);
  req.send(null);
}

function showArray(){
  var ready=req.readyState;
  var httpStatus=req.status;
  if (ready==4){
    if (httpStatus==200 || httpStatus==0){
      peopleArray = eval(req.responseText);
      myVar = "";
      for (i=0;i < peopleArray.length;i++){
        myVar = myVar + peopleArray[i] + ' ';
      }
      alert(myVar);
    }
  }
}
</script>
```

The result is the same alert box as before, but the important thing to note is that the Javascript does not contain any dynamically generated code like the previous example. Instead, it calls the CFML template, which generates the JSON data dynamically and returns it to the Javascript. In other words, if we were to change the CFML template to execute a query and build an array from the query, the Javascript alert box would dutifully report the members of the new array when the HTML page is run.

Javascript to ColdFusion

The trip to carry complex data from Javascript back to ColdFusion is slightly more involved, and requires the use of non-CF resources. Since Javascript executes on the browser, ColdFusion doesn't have a chance to interact with Javascript data unless that data is sent to the CF server. So we need a way to serialize complex Javascript data into JSON. This is accomplished through the use of any of several JSON stringifiers, available through any search engine and typically named **json.js**.

To use the json.js library, its functions must be made available to the template. The following code takes care of it:

```
<script language="Javascript" src="json.js"></script>
```

After the library is loaded, we can use the JSONSerialize() function to turn Javascript data into a JSON packet (your chosen json.js will probably have a different function name):

```
<script language="Javascript">
  JSONPacket = JSONSerialize(aryRecords);
</script>
```

Finally, we need a way to send the JSON packet to the CF server. We can do this by storing the JSON packet as the value of a hidden form input and submitting the form:

```
<form name="myForm" action="processData.cfm" method="post">
  <input type="hidden" name="inJSON">
</form>

<script language="Javascript">
  document.myForm.inJSON.value = JSONPacket;
  document.myForm.submit();
</script>
```

Using this method, the JSON packet gets submitted to the processData.cfm template as a variable named form.inJSON. We can use the DeSerializeJSON() function to deserialize the JSON packet for use in the template:

```
<CFSET aryFromJavascript = DeSerializeJSON(form.inJSON)>
```

This completes the Javascript-to-ColdFusion loop.

Summary

Like WDDX, JSON provides a means of moving data from the ColdFusion (server) side of things to the Javascript (browser) side and vice versa. The major difference between the two is that JSON is actually executable Javascript object notation, so there's no need for a special library to deserialize ("destringify") JSON data in Javascript. We do, however, need a library to serialize ("stringify") JSON data in Javascript. There are many such libraries available on the Internet.

We've now had an in-depth look at the use of lists, arrays, and structures in ColdFusion, as well as an introduction to WDDX and JSON and data interchange with Javascript. The next chapter provides some ideas about where we might want to explore from here.

6.Loose Ends

Having finished our trek through ColdFusion's most commonly used complex data types, we should take a few minutes to look over where we've been and some areas for further exploration. This chapter offers a brief summary of each chapter's topic, and suggests other topics to help you on your way.

Lists

A list is just a string comprised of one or elements separated (delimited) by a particular character. The default delimiter is a comma, but you can use any delimiter you choose. Lists with delimiters other than a comma may be though of as covert lists, because they are usually not thought of as lists. An example of a covert list is a phone number in the form of 212-555-1212. The delimiter in this case is a hyphen, and the list elements are the area code, exchange, and number.

Lists are the simplest of the data types we look at in this book. They can be created in many ways, including ColdFusion's List functions and any of the techniques we normally use to manipulate strings.

Arrays

Arrays can be thought of as super-powered lists. Instead of using a delimiter character, arrays have built-in structure that separates the elements. Each element of an array is referenced by an index number surrounded by square brackets (like myArray[2]). An element of an array can hold any data type. Arrays are created using the ArrayNew() function.

Arrays can be nested; we call nested arrays multi-dimensional arrays. Members of multi-dimensional arrays are referenced by an index number surrounded by square brackets for each dimension (like myArray[2][3]).

Arrays are great for processes that need to manipulate an unknown number of elements. Looping over an array is one of the most common programming practices.

Arrays in ColdFusion cannot be stored in the client scope, but we can serialize arrays into WDDX packets for storage in the client scope.

Structures

Structures are known as associative arrays in some programming languages. A structure is comprised of elements (values) that are associated with identifiers (keys). Each element of a structure is referenced by its key preceded by dot (dot notation), for example myStructure.firstName. An element of a structure may also be referenced by it key represented as a string and surrounded by square brackets (bracket notation), for example myStructure["firstName"]. Since any expression that evaluates to a string is valid inside the brackets, bracket notation can be very useful when looping over a structure. Structures may be created using the StructNew() function, or by storing a value to an element using dot notation.

Structures may be nested. When structures are nested, the elements may be referenced using dot notation (myStructure.Products.Description) or bracketed notation (myStructure["Products"]["Description]").

Structures are used when each value to be stored has a unique key associated with it. A referenced structure element carries more information in code (Person.firstName) than a referenced array element (Person[1]).

Structures cannot be stored in the client scope, but we can serialize structures into WDDX packets for storage in the client scope.

WDDX

The Web Distributed Data Exchange (WDDX) standard specifies a way to convert (serialize) complex data into an XML-based representation of the data. This representation is called a WDDX packet.

Because it is just string data, a WDDX packet can be transferred to practically any computer system. Once in the new environment, the packet can be converted into a complex data type that makes sense to the receiving system (deserialized).

In ColdFusion, the two most common uses of WDDX are to store complex data in the client scope and to exchange complex data with Javascript.

JSON

Javascript Object Notation (JSON), like WDDX, is a means to encode complex data in text form. Compared with WDDX, JSON is a lightweight notation, requiring fewer text characters to encode the same data. JSON is often used within AJAX applications, due to its native Javascript characteristics.

ColdFusion directly supports JSON manipulation as of ColdFusion 8. Earlier versions require the use of CFML-based third-party JSON libraries.

Putting It All Together

So we've explored the wild world of lists, arrays and structures in CFML. As a final exercise, we're going to take a look at how the same task could be accomplished using each datatype, and compare the benefits and challenges of using each one for the chosen task.

The task I've chosen for this exercise is a grade book for a class of students. This is an example that should be readily recognized by most everybody. To do the comparison, we'll need to set up the class with students, record some grades, and calculate some averages.

Set Up

We can think of the class as a container that holds students. As such, each member of the container is one student. We'll set up the class as a list of students first.

```
<cfset classList = “Adam, Barbara, Charlie, Deborah, Edward”>
```

This is quite straightforward. The code is simple, and we can easily see everyone in the class. Now let's set up the class as an array. We'll start with a syntax that will work for all versions of CFML.

```
<cfset classArray = ArrayNew(1)>
<cfset ArrayAppend(classArray,"Adam")>
<cfset ArrayAppend(classArray,"Barbara")>
<cfset ArrayAppend(classArray,"Charlie")>
<cfset ArrayAppend(classArray,"Deborah")>
<cfset ArrayAppend(classArray,"Edward")>
```

As an aside, we could also use the classList as the basis for a loop to create the array:

```
<cfset classArray = ArrayNew(1)>
<cfloop list="#classList#" index="thisStudent">
 <cfset ArrayAppend(classArray, thisStudent)>
</cfloop>
```

If using ColdFusion 8, the syntax can become much more like the list syntax:

```
<cfset classArray = ["Mary","Barbara","Charlie","Deborah","Edward"]>
```

Finally, we'll set up the class as a structure. Again, we'll start with a syntax that will work for all versions of CFML.

```
<cfset classStruct = StructNew()>
<cfset classStruct.Adam = "">
<cfset classStruct.Barbara = "">
<cfset classStruct.Charlie = "">
<cfset classStruct.Deborah = "">
<cfset classStruct.Edward = "">
```

We could also loop over the classList to create a structure:

```
<cfset classStruct = StructNew()>
<cfloop list="#classList#" index="thisStudent">
 <cfset classStruct[thisStudent] = "">
</cfloop>
```

And we have the ColdFusion 8 short syntax as well:

```
<cfset classStruct = {Adam="", Barbara="", Charlie="", Deborah="", Edward=""}>
```

The strengths and weaknesses of each approach don't really reveal themselves yet, because we haven't tried to do anything with the data yet. So on we go with the next task: adding scores for each student.

Add Grades

Using our first datatype, the list, we immediately have a problem. Each member of a list is a simple datatype, and we now need to store multiple values (name and one or more grade) for each student. How can we solve this?

One way might be to create an additional list for each graded assignment. Then each student would have a member in that list to store the grade. So if we have two assignments, the code might look like this:

```
<cfset assignment01List = "84,77,100,96,92">
<cfset assignment02List = "94,87,100,98,98">
```

This gets the job done, but it has some obvious drawbacks. First, the data for each student is stored in several places. If we forget one of those places in the calculations we'll need to do later, our results will be wrong. Second, there's nothing in the assignment lists to identify which student received which grade, other than the order of the list. If we lose an element in the list, or in any way reorder either the class list or any assignment list, the data will be corrupted.

How else might we solve this problem with a list? We could use a compound list, where each element of the primary list is another list. Here's how the data shown above would be expressed in this way (I've chosen the colon as the delimiter for the sublists):

```
<cfset classList = "Adam:84:94,Barbara:77:87,Charlie:100:100,Deborah:96:98,Edward:92:98">
```

This approach keeps all the data for a student associated with the student. It might present some difficulties in calculation later, but we'll wait to see how that works in the coming sections.

Now let's look at capturing scores in an array. We've already decided that, even though we could create a separate array for each assignment, that's not a terribly great idea. So we'll create a multi-dimensional array to store the class and their grades. We'll work from the basis of the classArray we already created. For each student, we need to change the value of the member from a string into another dimension and add the name and scores in the second dimension, so we'll loop over the array and do all that. (We're cheating a bit and using the assignment lists from the previous section to get the grades.)

```
<cfloop from="1" to="#ArrayLen(classArray)#" index="student">
  <cfset temp = classArray[student]>
  <cfset classArray[student] = ArrayNew(1)>
  <cfset classArray[student][1] = temp>
  <cfset classArray[student][2] = ListGetAt(assignment01List,student)>
  <cfset classArray[student][3] = ListGetAt(assignment02List,student)>
</cfloop>
```

The results of this operation are shown below. We can clearly see that this two-dimensional array is an array of arrays.

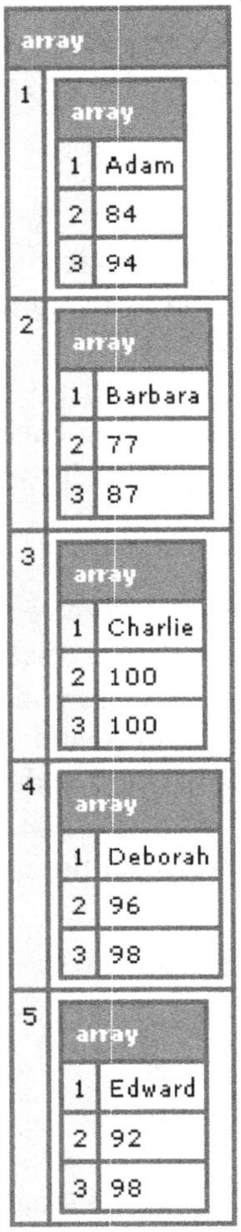

Illustration 52: classArray with Grades

Now let's add the grades to our class structure. Once again, we're going to use the previously defined lists so the code won't be pages long.

```
<cfset studentNum = 1>
<cfloop list="#classList#" index="thisStudent">
  <cfset grades = StructGet("classStruct.#thisStudent#.grades")>
  <cfset grades.homework = ListGetAt(assignment01List,studentNum)>
  <cfset grades.test = ListGetAt(assignment02List,studentNum)>
  <cfset studentNum = studentNum + 1>
</cfloop>
```

This results in a structure that is somewhat similar to the array in the previous illustration, with the notable difference that the grades have their own keys, **homework** and **test**.. The structure is shown here:

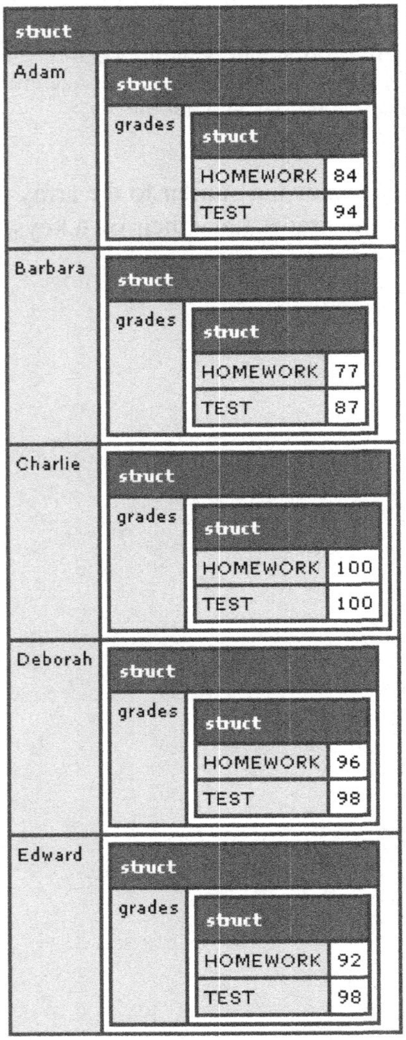

Illustration 53: classStruct with Grades

We've seen ways to store class and grade data in a list, an array, and a structure. But having the data isn't the same as using the data. The next couple of sections will look at how we would need to manipulate each datatype to do some common tasks.

Add Another Grade

Unless we're setting up our little application at the end of the school year, it's reasonable to assume we'll need to add individual grades as the year goes along. Let's take a look at what that entails for each of our datatypes.

With the compound list we've developed, we first need to find the student we need, then add the grade to that student's sublist. So let's say we need to add a grade of 92 to Adam's set of grades. Here's the code:

```
<cfset studentName="Adam">
<cfset gradeToAdd=92>

<cfset thisPosition = 1>
<cfloop list="#classList#" index="thisStudentList">
  <cfif ListFirst(thisStudentList,":") EQ studentName>
    <cfset studentList = thisStudentList>
    <cfset studentList = ListAppend(studentList,gradeToAdd,":")>
    <cfset classList = ListSetAt(classList,thisPosition,studentList)>
  </cfif>
  <cfset thisPosition = thisPosition + 1>
</cfloop>
```

The result of this is the list
Adam:84:94:92,Barbara:77:87,Charlie:100:100,Deborah:96:98,Edward:92:98

Note that Adam now has one more grade than any other student. We still don't know what each assignment belongs to each grade, but that's the limitation of using this list design. The code to actually do the task is a bit convoluted. We need to loop over the class list, check each sublist for the desired student, then extract that student's sublist, add a value to it, and put it back into the class list. Since we're looping over the list, we also need to keep track of the position within the list in order to be able to do the update with the ListSetAt() function. It works, but it can't exactly be called an elegant solution.

Now we'll do the same add-a-grade task with our class array:

```
<cfset studentName="Adam">
<cfset gradeToAdd=92>

<cfloop from="1" to="#ArrayLen(classArray)#" index="i">
  <cfif classArray[i][1] EQ studentName>
    <cfset ArrayAppend(classArray[i],gradeToAdd)>
  </cfif>
</cfloop>
```

After running this code, the classArray looks like this:

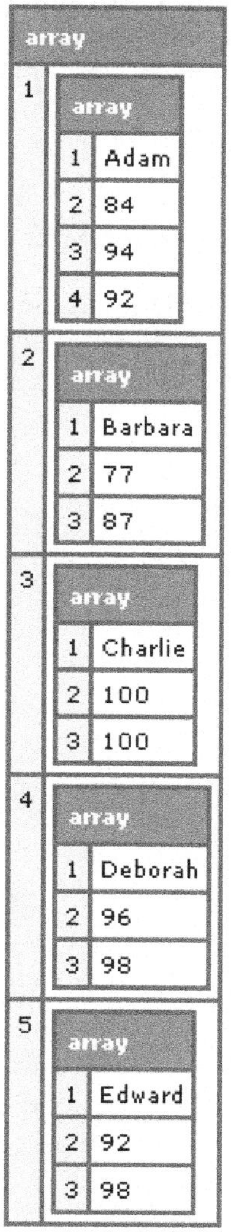

Illustration 54: classArray After Adding a Grade

So Adam has his new grade, and the code was quite a bit simpler than the list code. We still don't know what the assignment is for each grade. There are ways to fix that shortcoming while sticking with this array design, but I don't recommend them, so we'll go on to our structure design and see how that works out.

We need one more piece of data for the task now: the fact that this latest grade is for a project assignment. Here's the code for adding Adam's grade to the class structure:

```
<cfset studentName="Adam">
<cfset gradeToAdd=92>
<cfset assignment = "project">

<cfset classStruct[studentName].grades[assignment] = gradeToAdd>
```

Yes, that's all there is to it. One line of code to do the whole job. The illustration shows the result.

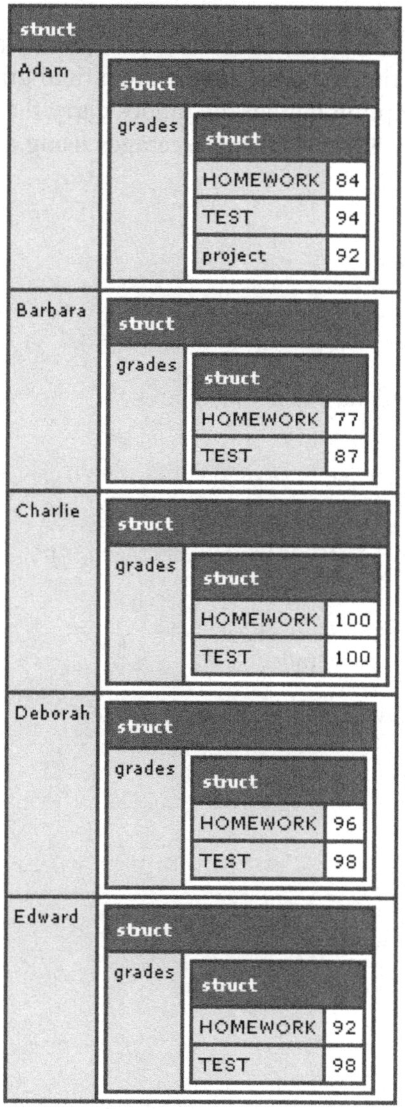

Illustration 55: classStruct After Adding a Grade

Calculate Averages

We've now explored some maintenance issues with our class records. Now let's look at what we would need to do when the time comes to calculate a final grade for each student. We'll make the highly unlikely assumption that all the grades carry the same weighting factor, again for the sake of simplicity. First we'll find averages using the class list:

```
<table>
 <tr>
  <th>Name</th><th>Average</th>
 </tr>
<cfloop list="#classList#" index="thisStudentList">
 <tr>
 <cfset gradeTotal = 0>
 <cfloop from="1" to="#ListLen(thisStudentList,':')#" index="i">
  <cfif i EQ 1>
   <td>
   <cfoutput>#ListGetAt(thisStudentList,i,":")#</cfoutput>
   </td>
  <cfelse>
   <cfset gradeTotal = gradeTotal + ListGetAt(thisStudentList,i,":")>
  </cfif>
 </cfloop>
 <cfset numberOfGrades = ListLen(thisStudentList,':')-1>
 <cfset average = gradeTotal/numberOfGrades>
 <td style="text-align:right;">
 <cfoutput>#NumberFormat(average,"99.99")#<br></cfoutput>
 <td>
 </tr>
</cfloop>
</table>
```

The resulting table is shown here:

Name	Average
Adam	90.00
Barbara	82.00
Charlie	100.00
Deborah	97.00
Edward	95.00

Illustration 56: Class Averages

Just like adding data, getting results from the list-based model required some looping logic and bit of careful planning for the code. Let's see what we need to do with the array-based model to get the same output.

```
<table>
  <tr>
    <th>Name</th><th>Average</th>
  </tr>
<cfloop from="1" to="#ArrayLen(classArray)#" index="i">
  <tr>
        <cfset gradeTotal = 0>
        <cfset thisStudentArray = Duplicate(classArray[i])>
        <cfset ArrayDeleteAt(thisStudentArray, 1)>
        <cfset gradeTotal = ArraySum(thisStudentArray)>
        <cfset numberOfGrades = ArrayLen(thisStudentArray)>
        <cfset average = gradeTotal/numberOfGrades>
    <td>
        <cfoutput>#classArray[i][1]#</cfoutput>
    </td>
      <td style="text-align:right;">
        <cfoutput>#NumberFormat(average,"99.99")#<br></cfoutput>
      <td>
  </tr>
</cfloop>
</table>
```

Notice that we don't need to use a nested loop for the array-based model. This is because we can use the ArraySum() and ArrayLen() functions to do the math for us. We use the

Duplicate() function to copy each student's subarray so we can delete the first element (the student's name) without disturbing the original array. Then we're ready to use ArraySum() on the new, numeric-only copy of the subarray.

So let's see how to accomplish the same thing with our structure:

```
<table>
  <tr>
    <th>Name</th><th>Average</th>
  </tr>
<cfloop collection="#classStruct#" item="thisStudent">
  <tr>
  <cfset gradeTotal = 0>
  <cfset numberOfGrades = 0>
  <cfloop collection="#classStruct[thisStudent].grades#" item="thisGrade">
      <cfset gradeTotal = gradeTotal + classStruct[thisStudent].grades[thisGrade]>
      <cfset numberOfGrades = numberOfGrades + 1>
  </cfloop>
  <cfset average = gradeTotal/numberOfGrades>
  <td>
    <cfoutput>#thisStudent#</cfoutput>
  </td>
  <td style="text-align:right;">
  <cfoutput>#NumberFormat(average,"99.99")#<br></cfoutput>
  <td>
  </tr>
</cfloop>
</table>
```

So we're back to using a nested loop, but it's a bit simpler to read, and the code tends to be much more self-explanatory, thanks to the existence of sensible keys within the structures.

It's important to recognize that the code presented here is by no means the only way to get the results we wanted. You may well have thought of other approaches yourself as you read through the examples. Each datatype has its own characteristics, strengths and weaknesses. While they are all quite flexible, a given programming task will probably lend itself more to one type than the others. Here are some characteristics to keep in mind:

Lists – can contain only simple datatypes (string, numeric, boolean); have implicit indices in terms of the element's position in the list; multiple "dimensions" require a contorted use of sublists with alternate delimiters

Arrays – can contain any datatype; have explicit numeric indices for members; can have multiple dimensions (nested arrays); don't have descriptive keys for members; have nice numeric manipulation functions in CF like ArraySum(), ArrayMin() and ArrayMax().

Structures – can contain any datatype; have descriptive keys for members; can be nested; don't have numeric manipulation functions; values can be directly referenced through use of keys.

> NOTE: ColdFusion 8 also brings some new operators into the game. For example, in the previous examples, any of the incrementing statements could have used the new ++ incrementing operator, e.g. <cfset numberOfGrades ++>

Where Do We Go From Here?

Once you're comfortable with manipulating complex data, your curiosity about similar subjects will probably get fired up at some point. There are other ColdFusion-related topics and some more generic topics that are natural extensions of the material in this book. In particular, you might want to look into queries (recordsets) as a data type in ColdFusion, ColdFusion Components (CFCs), XML and the XML Document data type in CFMX, and Fusebox.

Queries/Recordsets

Queries (also referred to as recordsets) can be thought of as a complex data type in ColdFusion. We create them with the <CFQUERY> tag or build them with recordset manipulation tags, and we can loop over them, referencing elements in ways that resemble structure manipulation at times, and array manipulation at other times. (For some insight into the recordset manipulation tags, see the <CF_QUERYSIM> custom tag, available at halhelms.com.)

While most every CF developer has used <CFQUERY> quite a lot, comparatively few really explore the various ways you can manipulate the results. Did you know you can easily extract all the values from a single column of a query? There are many such capabilities that should be explored if you want to have a complete CF toolkit at your service.

ColdFusion Components (CFCs)

ColdFusion Components were introduced in ColdFusion MX. Put very simplistially, a CFC is a structure with processes (called methods) attached. CFCs are a very friendly step toward the world of object orientation, and can bring a great deal of organization to your application development.

You can discover CFCs in the Techspedition book Discovering CFCs (ColdFusion MX Components) by Hal Helms and Ben Edwards.

XML Documents

XML by itself is an important technology that deserves your attention as a developer. I've often said that XML is the reason ASCII was invented lo these many years ago.

Specific to ColdFusion, you'll want to investigate the XML Document data type (introduced in ColdFusion MX). An XML document behaves in general like a nested structure, but it has some array-like characteristics as well. Being aware of how XML documents behave in ColdFusion and the attributes they have is an invaluable skill set.

You can start exploring XML documents with a copy of ColdFusion MX and the `<CFDUMP>` tag. Explore the `parseXML()` function to create an XML document variable, then examine it with `<CFDUMP>`. Look at the whole document, and at its elements. You can also read about this data type in the online documentation included with ColdFusion MX.

Fusebox

You may be aware of my involvement with the Fusebox community. The Fusebox core files make extensive use of lists, arrays, and structures; you might be interested in looking at what goes on in there.

With respect to XML and complex data, Fusebox recommends the use of Fusedoc, an XML-based documentation system. Fusedoc is very powerful, in that it lets us use tools like my Harness2, a utility that reads Fusedocs and creates test harness files for all the code in a Fusebox application. Of course, Harness2 makes extensive use of complex data types; you might enjoy looking at how it works, too. All my free code is available at http://www.grokfusebox.com.

You can get information about Fusebox from http://www.fusebox.org. You can pick up Harness2 and other Fusebox-related information at my site, http://www.grokfusebox.com.

There are also several great books available on Fusebox:

Fusebox 3

Fusebox: Developing ColdFusion Applications, by Jeff Peters and Nat Papovich

Discovering Fusebox 3 with ColdFusion, by Hal Helms and John Quarto-vonTivadar

Fusebox 4

Fusebox 4: Discovering Fusebox 4, by John Quarto-vonTivadar et al.

Fusebox 4 & FLiP: Master-Class ColdFusion Applications, by Jeff Peters

Fusebox 5

Fusebox 5 & FLiP: Master-Class ColdFusion Applications, by Jeff Peters

Those are all the loose ends I can think of for the moment. If you think of others that should be addressed, drop me a line at jeff@grokfusebox.com. I hope you enjoy working with your newly honed ColdFusion tools.

7.Appendix A - List Functions Quick Reference

ListAppend(list, newMember [, delimiter])

ListChangeDelims(list, newDelimiter)

ListContains(list, targetString [, delimiters])

ListContainsNoCase(list, targetString [, delimiters])

ListDeleteAt(list, position [, delimiters])

ListFind(list, targetString [, delimiters])

ListFindNoCase(list, targetString [, delimiters])

ListFirst(list [, delimiters])

ListGetAt(list, position [, delimiters])

ListInsertAt(list, position, newMember [, delimiters])

ListLast(list [, delimiters])

ListLen(list [, delimiters])

ListPrepend(list, newMember [, delimiters])

ListQualify(list, qualifyer [, delimiters] [,"CHAR"|"ALL"])

ListRest(list [, delimiters])

ListSetAt(list, position,value [, delimiters])

ListSort(list [, "numeric"|"text"|"textnocase"] [, "asc"|"desc"] [, delimiters])

ListToArray(list [, delimiters])

ListValueCount(list, value [, delimiters])

ListValueCountNoCase(list, value [, delimiters])

ReplaceList(listToProcess, membersToReplace, valuesToInsert)

8. Appendix B – Array Functions Quick Reference

ArrayAppend(array, newMember)

ArrayAvg(array)

ArrayClear(array)

ArrayDeleteAt(array, position)

ArrayInsertAt(array, position, value)

ArrayIsEmpty(array)

ArrayLen(array)

ArrayMax(arrayWithNumericValues)

ArrayMin(arrayWithNumericValues)

ArrayNew(1|2|3)

ArrayPrepend(array, newMember)

ArrayResize(array, newLength)

ArraySet(array, startPosition, endPosition, value)

ArraySort(array, "numeric"|"text"|"textnocase" [, "asc"|"desc"])

ArraySum(arrayWithNumericValues)

ArraySwap(array, positionA, positionB)

ArrayToList(array [, delimiter])

IsArray(variableName [, lengthToCompare])

ListToArray(list [, delimiter])

9.Appendix C - Structure Functions Quick Reference

IsStruct(variable)

StructAppend(donorStructure, recipientStructure, overwriteBoolean)

StructClear(structure)

StructCopy(structure)

StructCount(structure)

StructDelete(structure, key [, notifyIfNoKeyPresentBoolean])

StructFind(structure, key)

StructFindKey(structure, keyToFind, "ONE"|"ALL")

StructFindValue(structure, value)

StructGet(structureName)

StructInsert(structure, key, value [, overwriteBoolean])

StructIsEmpty(structure)

StructKeyArray(structure)

StructKeyExists(structure, key)

StructKeyList(structure [, delimiter])

StructNew()

StructSort(structure [, "TEXT"|"NUMERIC"][, "ASC"|"DESC"][, pathToSubElement])

StructUpdate(structure, key, valueToAssign)

10.Appendix D – New ColdFusion 8 Functions Quick Reference

DeSerializeJSON((variable)

DeSerializeWDDX((variable)

IsJSON(variable)

IsWDDX(variable)

SerializeJSON(variable)

SerializeWDDX(variable)

Index

Made in the USA
Monee, IL
05 November 2026

67378312R00098